ELECTRICAL ENGI~~~~~~~

PHASE-MODE TRANSFORMATION MATRIX APPLICATION FOR TRANSMISSION LINE AND ELECTROMAGNETIC TRANSIENT ANALYSES

ELECTRICAL ENGINEERING DEVELOPMENTS

Additional books in this series can be found on Nova's website under the Series tab.

Additional E-books in this series can be found on Nova's website under the E-book tab.

ELECTRICAL ENGINEERING DEVELOPMENTS

PHASE-MODE TRANSFORMATION MATRIX APPLICATION FOR TRANSMISSION LINE AND ELECTROMAGNETIC TRANSIENT ANALYSES

AFONSO JOSÉ DO PRADO

SÉRGIO KUROKAWA

LUIZ FERNANDO BOVOLATO

JOSÉ PISSOLATO FILHO

AND

EDUARDO COELHO MARQUES DA COSTA

Nova Science Publishers, Inc.

New York

For permission to use material from this book please contact us:
Telephone 631-231-7269; Fax 631-231-8175
Web Site: http://www.novapublishers.com

NOTICE TO THE READER

LIBRARY OF CONGRESS CATALOGING-IN-PUBLICATION DATA

Phase-mode transformation matrix application for transmission line and electromagnetic transient analyses / authors, Afonso Josi do Prado ... [et al.].
    p. cm.
  Includes bibliographical references and index.
  ISBN 978-1-61728-486-1 (softcover)
  1. Transients (Electricity) 2. Electric power distribution--Alternating current. 3. Phase distortion (Electronics) I. Prado, Afonso Josi do.
  TK3226.P496 2010
  621.319'21--dc22
                        2010044741

Published by Nova Science Publishers, Inc. † New York

CONTENTS

PREFACE

Frequency-dependent transmission line models can be developed in phase domain or in mode domain. Associated to these transmission line models, time domain or frequency domain can be used for electromagnetic transient analysis solutions. Because of these options, there are several models for transmission line representation in electromagnetic transient studies. One reason for this is the longitudinal parameter frequency dependence. This book presents a variety of models which serve to improve the frequency dependence representation. In some cases, phase-mode transformation is applied, considering the problem in mode domain, and searching for improving the frequency-dependent parameter representation. On the other hand, changes in the system structure as well as voltage and current surge waves are better modeled on the time domain. Hybrid models can be also used considering the advantages of the frequency and the time domains. These models are based on phase-mode transformations. Some applications and analyses of the mentioned phase-mode transformations are also carried out in this text.

INTRODUCTION

Frequency-dependent transmission line models can be developed in phase domain or in mode domain. Associated to these transmission line models, time domain or frequency domain can be used for electromagnetic transient analysis solutions [1-6]. Because of these options, there are several models for transmission line representation in electromagnetic transient studies [7-12]. One reason for this is the longitudinal parameter frequency dependence [3-6, 11-19]. With the aim to improve the frequency dependence representation, several models have been suggested [1-19]. In some cases, phase-mode transformation is applied, considering the problem in mode domain, and searching for improving the frequency-dependent parameter representation [3-6, 8-9]. On the other hand, changes in the system structure as well as voltage and current surge waves are better modeled on the time domain. So, a hybrid model based on phase-mode transformation can be applied using a better line parameter frequency-dependence representation with the time domain model advantages.

In exact mathematical development, the phase mode transformation matrices depend on the frequency because the line parameters are frequency dependent. Because of this, all values that represent the electrical line characteristics, such as the longitudinal impedance (Z) and transversal admittance (Y) matrices, are influenced by frequency [3, 11-19]. Using the exact phase-mode transformation, we calculated the line parameters. It is an alternative methodology to calculate transmission line parameters per unit length. With this methodology the transmission line parameters can be obtained from impedances measured in one terminal of the line. So, a new procedure is shown to calculate frequency-dependent transmission line parameters directly from currents and voltages from the line terminals.

If the line is a symmetric three-phase line, the phase-mode transformation can be carried out using a 3-order real and constant matrix and a 2-order frequency dependent transformation matrix. For analyses and simulations, the computational time can be decreased because the order of frequency dependent matrix is reduced. Also using the exact phase-mode transformation matrices and considering the frequency range mentioned above, an alternative procedure is carried out for the equivalent conductor determination of a bundle of subconductors. Considering symmetrical bundles, the proposed alternative obtains results that are similar to those obtained from the geometric mean radius (GMR) procedure. For asymmetrical bundles, the alternative procedure based on phase-mode transformations is more accurate when frequency values in the range from 10 Hz to 100 Hz are considered.

If the mathematical model based on the exact phase-mode transformation is applied to digital programs, the result can be a slow digital routine for transmission line analyses. An alternative that can be considered is the use of a single real transformation matrix. It is a way to obtain fast transmission line transient simulations as well as to avoid convolution procedures in this simulation type [4-6, 8-10]. Clarke's matrix has presented interesting performances for three-phase transmission lines: exact results for transposed cases and negligible errors for non-transposed ones [7]. This matrix is single, real, frequency independent, and identical for voltage and current. So, it is analyzed the changing the exact phase-mode transformation matrices into matrices composed of constant and real elements. The phase-mode transformation matrix applications can lead to complicated and slow numeric routines. An alternative is the single real matrix use for a wide frequency range. It is possible because the elements of the eigenvectors that are the phase-mode transformation matrices smoothly vary in function of frequency. In this case, Clarke's matrix is used and checked for symmetrical and asymmetrical three-phase transmission lines considering transposed and untransposed cases. For transposed cases, Clarke's matrix is an eigenvector of the three-phase transmission lines. For untransposed cases, this matrix is a good approximation to the exact transformation considering the eigenvalue comparisons. However, this is not true for off-diagonal elements of the matrix obtained from the application of Clarke's matrix. Using a correction procedure, the elements of Clarke's matrix are corrected leading to two new transformation matrices: one related to the voltage values and the other related to the current values. Because the obtained errors are negligible and the mentioned off-diagonal elements become negligible values, these new matrices can be considered eigenvectors of the analyzed lines. These matrices smoothly vary in function of frequency, and their elements have small imaginary part. They are checked using untransposed symmetrical and untransposed asymmetrical three-phase transmission lines.

Finally, for future development, the obtained numeric routines will be applied with state variables for electromagnetic transient simulations. So, it is suggested a model for transmission line that considers the frequency influence using state variables. Another suggestion for future development is searching adequate transformation matrices for systems with some parallel three-phase circuits.

Chapter 1

TRANSMISSION LINE PARAMETERS DERIVATION FROM IMPEDANCE VALUES [20]

The self and mutual impedances present in the overhead transmission line equations in the frequency domain can be derived from the solution of Maxwell's equations for the boundary conditions at the contact surfaces of the three relevant materials: conductor, air and ground. [21]. Evaluating line parameters, there are some usual assumptions that imply in physical approximations related to line geometry or electromagnetic field behavior [22]. The main line geometric approximations are: the soil surface is a plane; the line cables are horizontal and parallel among themselves; the distance between any pair of conductors is much higher than the sum of their radii; the electromagnetic effects of structures and insulators are neglected. About electromagnetic field behavior for line transversal parameters, it is assumed the quasi-stationary electromagnetic field simplification [22]. The most used procedures that represent the ground assume a constant and frequency independent ground conductivity, neglecting the ground dielectric permittivity. There are situations where the mentioned simplifications can not be assumed. An example of a situation in which physical properties can be changed is in the study of electromagnetic transients that include nonuniform lines that could be portions of transmission lines where the conductors are not parallel [23].

The transmission line parameters could be derived from measured transmission line impedances, but there are some practical difficulties to measure the frequency response of a transmission line [24]. It requires a strong motivation to get the consent from a power utility to make an outage of a long transmission line. The experimental setup may be nontrivial. Frequency measurement requires a voltage source with variable frequency and high power. Due to difficulties above mentioned the procedure developed in this chapter has been used in transmission lines represented by digital models. So, it proposes a methodology to calculate longitudinal and shunt parameters per unit length of overhead transmission lines from the impedances measured at line terminal.

I.1. FREQUENCY-DEPENDENT TRANSMISSION LINE PARAMETER CALCULATING

One of the most important aspects for modeling of electromagnetic transient in transmission lines is that the line parameters depend on the frequency. Models that assume constant parameters have not adequately simulated the response of the line over the wide range of frequency for transient conditions. The constant line parameter representation produces a magnification of the higher harmonics, a general distortion of the wave shapes and exaggerated magnitude peaks [17].

Using the frequency domain, the self and mutual impedances included in the overhead transmission line equations can be derived from the solution of Maxwell's equations. In these equations, the impedance matrix (Z) determination falls into three parts: (self impedance) or two parts (mutual impedance): internal longitudinal, external longitudinal and soil effect impedances. The internal longitudinal impedance is associated with the electromagnetic field within the conductor. Electromagnetic fields do not affect mutual terms and the mutual impedance does not depend on the internal longitudinal impedance. Due to the skin effect, the resistance increases whereas the inductance decreases. The external longitudinal impedance is associated with the electromagnetic field outside the conductors. In this case, it is assumed a lossless ground and the other assumptions indicated before. For a lossy ground, it is considered the soil effect that means an additional parcel of the external longitudinal impedance matrix. A similar analysis applies also to transversal admittance matrix (Y) where, for typical conditions, it is reasonable to assume ideal conductors and ground up to 1 MHz.

The parameters of transmission lines with ground return are highly dependent on the frequency. Formulas to calculate the influence of the ground return were developed by Carson and Pollaczek and these formulas can also be used for power lines. Both lead to identical results for overhead lines, but

Pollaczek's formula is more general inasmuch as it can also be used for underground conductors or pipes [1, 2].

I.2. LINE PARAMETER CALCULATING FROM PHASE CURRENT AND VOLTAGE VALUES

The basic equations of a transmission line for sinusoidal alternated electrical magnitude complex representation are [22]:

$$\frac{d^2 V_{PH}}{dx^2} = Z \cdot Y \cdot V_{PH} \quad and \quad \frac{d^2 I_{PH}}{dx^2} = Y \cdot Z \cdot I_{PH} \tag{1}$$

The Z and Y matrices are per unit length longitudinal impedance and shunt admittance matrices, respectively. The elements of these matrices are frequency dependent. The V_{PH} and I_{PH} vectors are, respectively, transversal line voltage and longitudinal line current vectors. These equations are valid if the electromagnetic field has a quasi-stationary behavior in orthogonal direction to line axis [22].

Poly-phase transmission line equations can be solved transforming n coupled equations into n decoupled equations. Decoupling of equations is carried out using a suitable chosen modal transformation matrix T_I changing the YZ matricial product into its diagonal form [5, 11]:

$$T_I^{-1} \cdot Y \cdot Z \cdot T_I = \lambda \tag{2}$$

The λ matrix is the eigenvalue one.

Substituting equation (2) in equation (1), it is obtained the basic equations of a transmission line in mode domain [14]:

$$\frac{d^2 V_M}{dx^2} = Z_M \cdot Y_M \cdot V_M \quad and \quad \frac{d^2 I_M}{dx^2} = Y_M \cdot Z_M \cdot I_M \tag{3}$$

The Z_M and Y_M matrices are described as following [11]:

$$Z_M = T_I^T \cdot Z \cdot T_I \quad and \quad Y_M = T_I^{-1} \cdot Y \cdot T_I^{-T} \tag{4}$$

The T index identifies the transposition of the analyzed matrix. The negative T index is related to the inverse transposed matrix. The T_I^{-1} is the inverse T matrix. The V_M and I_M vectors are, respectively, transversal line voltages and longitudinal line currents in mode domain.

Because matrices Z_M and Y_M are diagonal matrices, the $Z_M Y_M$ and $Y_M Z_M$ matricial products are diagonal matrices and there are no couplings among modes. For a generic mode, it is carried out [14]:

$$\begin{cases} E_A = E_B \cdot \cosh(\gamma \cdot d) - I_B \cdot Z_C \cdot \sinh(\gamma \cdot d) \\ I_A = -I_B \cdot \cosh(\gamma \cdot d) + \dfrac{E_B}{Z_C} \cdot \sinh(\gamma \cdot d) \end{cases} \tag{5}$$

E_A and E_B are, respectively, modal voltages at terminals A and B in Figure I.1. The terms I_A and I_B are modal currents at these terminals and d is the line length. The terms γ and Z_C are, respectively, the propagation function and the characteristic impedance of the analyzed mode.

The Z_C and γ values are written as being [21]:

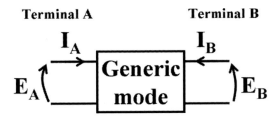

Figure I.1. Quadripole of a transmission line generic mode.

$$\gamma = \sqrt{Z_{KK} \cdot Y_{KK}}$$

$$Z_C = \sqrt{\dfrac{Z_{KK}}{Y_{KK}}} \tag{6}$$

In this case, Z_{KK} is an element of the Z matrix that represents the impedance of the k–mode. Y_{KK} is an element of the Y matrix that represents the impedance of the k-mode. The k–mode is the generic mode mentioned in Figure I.1.

Supposed a Z_L impedance load in terminal B shown in C Figure I.1. The Z_L value is known and, using equation (5), it is possible to determine the Z_{EC} impedance of the generic mode equivalent circuit [23]. The Z_{EC} is described as:

$$Z_{EC} = \frac{Z_L \cdot \cosh(\gamma \cdot d) + Z_C \cdot \sinh(\gamma \cdot d)}{\cosh(\gamma \cdot d) + \dfrac{Z_L}{Z_C} \cdot \sinh(\gamma \cdot d)} \tag{7}$$

Considering two specific line configurations, there are two impedances for the mode shown in Figure I.1. The first impedance is defined considering that terminal B is opened and the other impedance is defined considering terminal B is short-circuited. Z_{OPEN} is the equivalent impedance when terminal B is open ($Z_L \rightarrow \infty$), E_{AOPEN} is the voltage in terminal A when terminal B is open. I_{AOPEN} is the current related to this described line configuration. Manipulating equation (7), it is possible to write each Z_{OPEN} mode as function of γ e Z_C as following:

$$Z_L \rightarrow \infty \Rightarrow Z_{EC} = Z_{OPEN} = \frac{E_{AOPEN}}{I_{AOPEN}} = Z_C \cdot \tanh(\gamma \cdot d) \tag{8}$$

Using a similar manipulating, it is obtained:

$$Z_L = 0 \Rightarrow Z_{EC} = Z_{CC} = \frac{E_{ACC}}{I_{ACC}} = Z_C \cdot \coth(\gamma \cdot d) \tag{9}$$

In this case, Z_{CC} is the equivalent impedance when terminal B is short-circuited. For short-circuit, Z_L is null. E_{ACC} and I_{ACC} are, respectively, the voltage and current values in terminal A for this line configuration.

The Z_{OPEN} and Z_{CC} equivalent impedances can be calculated directly from currents and voltages of the line or directly from γ e Z_C. Using γ and Z_C related to Z_{CC} and Z_{OPEN} values, it is possible to calculate longitudinal and transversal transmission line parameters from equation (6).

It is considered a polyphase transmission line (n phases) where the sending ending terminal is called terminal A and the receipting ending terminal is called terminal B.. It is also considered that it is possible to obtain in these line terminals the V_{PH}, $I_{PH\ OPEN}$ and $I_{PH\ CC}$ vectors. The V_{PH} vector is composed by the sources connected in terminal A of the line. The $I_{PH\ OPEN}$ and

$I_{PH\,CC}$ vectors have the longitudinal currents in terminal A of each phase of the line, considering terminal B opened and short-circuited, respectively. The mentioned vectors are written in modal domain as being:

$$E_M = T_I^T \cdot V_{PH}$$
$$I_{M\,OPEN} = T_I^{-1} \cdot I_{PH\,OPEN}$$
$$I_{M\,CC} = T_I^{-1} \cdot I_{PH\,CC}$$

(10)

E_M is the vector with modal voltage sources. Each modal voltage is connected to terminal A of the respective mode of the line. The $I_{M\,OPEN}$ and $I_{M\,CC}$ vectors are current vectors in terminal A of each mode considering that terminal B is open and short-circuited, respectively. Considering n modes of a polyphase line, the equations (8) and (9) are used for calculating the equivalent impedances Z_{OPEN} and Z_{CC} of each mode. After that, manipulating this equation, it is obtained:

$$\coth(\gamma \cdot d) = \sqrt{\frac{Z_{OPEN}}{Z_{CC}}}$$

(11)

The Z_{OPEN} and Z_{CC} equivalent impedances are known and obtained from equations (8) and (9), respectively. Manipulation the last equation, it is obtained the propagation function (γ) of a generic mode of the line.

On the other hand, the last equation can be write as following:

$$\coth(\gamma \cdot d) = \frac{e^{\gamma \cdot d} + e^{-\gamma \cdot d}}{e^{\gamma \cdot d} - e^{-\gamma \cdot d}}$$

(12)

Equaling the equations (11) and (12), it is obtained:

$$\frac{e^{\gamma \cdot d} + e^{-\gamma \cdot d}}{e^{\gamma \cdot d} - e^{-\gamma \cdot d}} = \sqrt{\frac{Z_{OPEN}}{Z_{CC}}}$$

(13)

From equation (13), it is possible to express γ as:

$$\gamma = \frac{\ln(F_1) + j \cdot \arccos(F_2)}{2d}$$

(13)

F_1 and F_2 are determined by:

$$F_1 = \frac{(1+D_1)^2 + D_2^2}{\sqrt{(1-D_1^2-D_2^2)^2 + 4D_2^2}}$$

(14)

$$F_2 = \frac{D_1^2 + D_2^2 - 1}{\sqrt{(1-D_1^2-D_2^2)^2 + 4D_2^2}}$$

The D_1 and D_2 values are related to Z_{OPEN} and Z_{CC} as following:

$$\sqrt{\frac{Z_{OPEN}}{Z_{CC}}} = D_1 + j \cdot D_2$$

(15)

For determination of D_1 and D_2 values, equations (13) and (15) can be used in a specific frequency value applying the following restriction:

$$D_1 \neq \pm 1 \quad and \quad D_2 \neq 0$$

(16)

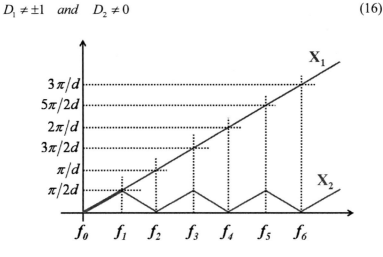

Figure I.2. The γ function imaginary part for a generic mode.

The last restriction is necessary for mode of the analyzed line in each specific frequency used for the γ calculating.

However, it can be observed that there is a no negligible difference between the imaginary part of the equation (13) and the correspondent value obtained from the equation (6). It is analyzed from the next equations, considering a generic mode.

$$X_1 = \mathrm{Im}\left(\sqrt{Z_{KK} \cdot Y_{KK}}\right)$$

$$X_2 = \frac{\arccos(F_2)}{2d}$$

(17)

The next graphic compares the X_1 and X_2 values that depend on the frequency.

Functions X_1 and X_2 are equal for frequencies between f_0 and f_1 only. For the other frequency ranges in the used graphic, there are increasing differences between the both functions. However, it is possible to equal X_1 and X_2 for other frequency ranges using a single procedure. Consider the H function as shown in the next equation.

$$H(f_K) = \left|\frac{\partial X_2(f_K)}{\partial f}\right|(f_K - f_{K-1}) + H(f_{K-1})$$

(18)

In this case, it can be used the following approximation:

$$\frac{\partial X_2(f_K)}{\partial f} \approx \frac{X_2(f_K) - X_2(f_{K-1})}{(f_K - f_{K-1})}$$

(19)

Equations (18) and (19) can be used if f_K is higher than f_1. If f_K is lower than f_1, $H(f)$ is written as following:

$$H(f_K) = X_2(f_K)$$

(20)

Using equations (18), (19) and (20), it is obtained a function relating X_1 and X_2 for each frequency.

Substituing the equation (6) in product of (8) and (9), it is obtained the following:

$$\frac{Z_{KK}}{Y_{KK}} = Z_{CC} \cdot Z_{OPEN}$$

(21)

Using equations (6) and (21), it is determined:

$$Z_{KK} = \gamma \sqrt{Z_{CC} \cdot Z_{OPEN}}$$

$$\text{(22)}$$

$$Y_{KK} = \frac{\gamma}{\sqrt{Z_{CC} \cdot Z_{OPEN}}}$$

The terms Z_{OPEN} and Z_{CC} are calculated from currents and voltages obtained in one terminal of a generic mode considering that the other terminal is open and short-circuited, respectively. Applying the proposed development, the modal propagation function of each mode is calculated, considering a frequency range of interest. From the modal propagation function values, longitudinal impedance and transversal admittance of each mode are derived. With longitudinal impedance and transversal admittance of each mode, longitudinal impedance matrix Z_M and transversal admittance matrix Y_M in mode domain are determined. These matrices are converted to the phase domain as following:

$$Z = T_I^{-T} \cdot Z_M \cdot T_I^{-1}$$

$$\text{(23)}$$

$$Y = T_I \cdot Y_M \cdot T_I^{-T}$$

The Z and Y matrices have been defined in equation (1) as the longitudinal impedance and the shunt admittance ones in phase domain, respectively. So, if currents and voltages for the opened and short-circuited line are known, the Z and Y matrices can be calculated.

The procedure proposed in this paper is based in the knowledge of the modal transformation matrix and the voltage and current magnitudes at the beginning of the line during open and short circuit conditions. These conditions can reduce the applicability of the methodology, if it is necessary to consider the exact values. If it is applied some approximations, the proposed procedure can be used obtaining adequate results for practical applications. The main restriction is that the modal transformation matrix should be known. For two-phase transmission line that actual samples are the transmission lines in continuous current, the modal transformation matrices are known for any frequency because these matrices are not influenced by the line geometric characteristics and the frequency.

For transposed three-phase transmission lines, considering symmetrical and asymmetrical cases, Clarke's matrix is an eigenvector matrix for any frequency value and it is independent of the line geometric characteristics.

Considering untransposed three-phase transmission lines, there are some suggestions that can be applied. These are analyzed in the next chapters.

Chapter 2

AN ALTERNATIVE MODEL FOR EQUIVALENT CONDUCTOR DETERMINATION FROM BUNDLED CONDUCTORS

Connecting two or more sub-conductors, bundled conductors can be an efficient alternative to increase the capacity of high voltage transmission lines without conductor gauge increasing and maintaining the electromagnetic interference in acceptable levels. The bundled conductors are composed by sub-conductors connected in parallel and it is used spacers for attaching the bundle to the towers or along the sags among the line towers [25]. In actual systems, up to 230 kV, for the most of transmission lines, each phase is composed by a bundled conductor [26-28]. Using the mentioned spacers, the spacing among the adjacent sub-conductors is from 0.4 to 0.6 m for conventional lines. In these cases, the sub-conductors are equal and it can be considered that the current is equally distributed among the sub-conductors. These are the ideal conditions for the classical bundle conductor modeling application where the geometric mean radius (GMR) concept is applied. Using this concept, the bundle conductor is changed into an equivalent conductor at the geometric center of the bundle [29-35].

There are systems with non-symmetrical bundle conductors. There are also cases where the bundle conductor is composed by different sub-conductors. For these non conventional bundle conductor cases, the GMR concept is not appropriated and, because of this, an alternative procedure is introduced in this chapter, defining an equivalent conductor of a bundle conductor using the unbalanced distribution of the current among the sub-conductors.

II.1. ANALYSES FOR SINGLE CONDUCTORS

It is considered a simple system with to single conductors in Figure II.1. Based on this system, the analyses can be similarly carried out increasing the number of conductors.

The radius of conductor i is r_i, as well as, r_k is the radius of conducto k. The system shown in Figure II.1 can be characterized using the longitudinal impedance (Z) and the shunt admittance (Y) matrices per unit length. The Z matrix is written as:

$$Z = R + j\omega L \tag{24}$$

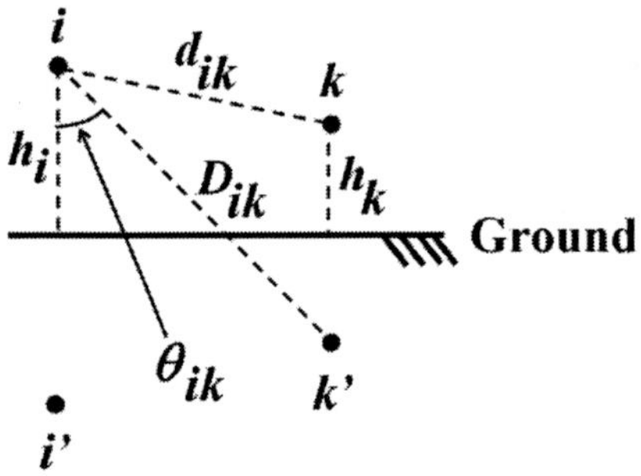

Figure II.1. System with two single conductors.

In the last equation, R is the longitudinal resistance matrix and L is the longitudinal inductance one. The R and L matrices are frequency dependent and also influenced by the skin and ground effects.

The Y real part matrix is the shunt conductance matrix (G) and the Y imaginary part matrix is the shunt capacitance matrix (C). For the most cases related to the power system conductors, the G matrix can be considered null and the C matrix can be considered frequency independent. The Y matrix is:

$$Y = G + j\omega C \cong j\omega C \tag{25}$$

Each self impedance (Z_{ii}), which is in the Z main diagonal, is written as adding three elements: the self external impedance ($Z_{ext\ ii}$), the self impedance due to the ground effect ($Z_{ground\ ii}$) and the self internal impedance ($Z_{int\ ii}$). Each mutual impedance (Z_{ik}), which is out of the Z matrix main diagonal, is written as adding two elements: the mutual external impedance ($Z_{ext\ ik}$) and the mutual impedance due to the ground effect ($Z_{ground\ ik}$). It is shown in the next equation.

$$Z_{ii} = Z_{ext\ ii} + Z_{groundii} + Z_{int\ ii}$$

$$Z_{ik} = Z_{ext\ ik} + Z_{groundik}$$

(26)

The external impedances are calculated as demonstrated in the next equations written for the i conductor of Figure II.1. In this case, d_{ik} is the distance between the i and k conductors and D_{ik} is the distance between the i conductor and the k conductor image.

$$Z_{ext\ ii} = j\omega\frac{\mu_0}{2\pi}\ln\left(\frac{2h_i}{r_i}\right) \quad and \quad Z_{ext\ ik} = j\omega\frac{\mu_0}{2\pi}\ln\left(\frac{D_{ik}}{d_{ik}}\right) \tag{27}$$

The ground effects for impedances are calculated using the next equations that are also written basing on the i conductor. Carson's infinite series are applied for determination of ground effect impedances.

$$Z_{groundii} = R_{ii}\left(a_{ii}, \varphi_{ii}\right) + j\omega L_{ii}\left(a_{ii}, \varphi_{ii}\right)$$

$$Z_{groundik} = R_{ik}\left(a_{ik}, \varphi_{ik}\right) + j\omega L_{ik}\left(a_{ik}, \varphi_{ik}\right)$$

(28)

Based on Figure II.1, it is determined the following values:

$$a_{ii} = 4\pi\sqrt{5}\cdot 10^{-4}\cdot h_i\sqrt{\frac{\omega}{2\pi\rho}} \quad a_{ik} = 4\pi\sqrt{5}\cdot 10^{-4}\cdot D_{ik}\sqrt{\frac{\omega}{2\pi\rho}}$$

(29)

$$\varphi_{ii} = 0 \quad\quad\quad \varphi_{ik} = \theta_{ik}$$

Taking the i conductor as a sample, the self internal impedance is calculated basing on the radius of this conductor (r_i). The m value is determined by equation (31) depending on the permeability and resistivity of the analyzed conductor.

$$Z_{intii} = \frac{j\omega\mu_i}{2\pi r_i}\left(\frac{ber(mr_i) + j\,bei(mr_i)}{ber\,'(mr_i) + j\,bei\,'(mr_i)}\right) \tag{30}$$

The m value in the last equation is calculated using the next equation where μ_i is the permeability of the conductor $_i$ and ρ_i is the resistivity of this conductor.

$$m = \sqrt{\frac{\omega\mu_i}{\rho_i}} \tag{31}$$

Neglecting the real part of the shunt admittance matrix, this matrix is equal to the shunt capacitance matrix that is calculated as:

$$C = P^{-1} \tag{32}$$

The P matrix is the Maxwell's potential coefficient one and every element of this matrix is calculating as following:

$$P_{ii} = \frac{1}{2\pi\varepsilon_0}\ln\left(\frac{2h_i}{r_i}\right) \quad and \quad P_{ii} = \frac{1}{2\pi\varepsilon_0}\ln\left(\frac{D_{ik}}{d_{ik}}\right) \tag{33}$$

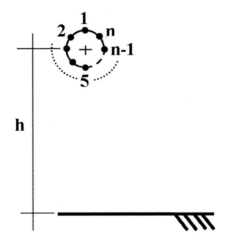

Figure II.2. A generic bundled conductor.

II.2. APPLICATION OF GEOMETRIC MEAN RADIUS (GMR) CONCEPT FOR CHANGING A BUNDLED CONDUCTOR INTO AN EQUIVALENT CONDUCTOR

A bundle conductor can be changed into an equivalent conductor. This equivalent conductor can be represented by the equivalent longitudinal impedance (Z_{EQ}) and the equivalent shunt admittance (Y_{EQ}). These values are determined using the same equations that are used for an individual conductor. So, the equivalent longitudinal impedance is composed by the external impedance, the impedance due to ground effect and the internal impedance. The shunt admittance is composed by the capacitance. It is shown in the next equation and based on Figure II.2 where h is the height of the equivalent conductor and corresponds to the height of the geometric center of the bundled conductor. It is considered n sub-conductors for this generic bundled conductor.

$$Z_{EQ} = Z_{EQ\ ext} + Z_{EQ\ ground} + Z_{EQ\ int}$$

$$(34)$$

$$Y_{EQ} = j\omega C_{EQ}$$

The equivalent external impedance is calculated as following:

$$Z_{EQ\ ext} = j\omega \frac{\mu_0}{2\pi} \ln\left(\frac{2h}{r_{GMR}}\right) \qquad (35)$$

In this case, h is the height of the symmetrical center of the bundled conductor and the r_{GMR} is the geometric mean radius (GMR) of the bundle. This last value is defined as:

$$r_{GMR} = \sqrt[n\cdot n]{\prod_{i=1}^{n}\left(r_i \cdot \prod_{k=1}^{n} d_{ik}\right)}, \quad k \neq i \qquad (36)$$

For the last equation, r_i is the i sub-conductor radius and d_{ik} is the distance between the sub-conductors i and k shown previously in Figure II.1.

Determining the equivalent ground impedance, the angle φ is null and the real and imaginary terms of the Z_{ground} value are based in the procedure used for individual conductors that is previously presented. It is represented as:

$$Z_{EQ\,ground} = R_{EQ}(a,\varphi) + j\omega L_{EQ}(a,\varphi) \tag{37}$$

Generally, the equivalent internal impedance is determined neglecting the shunt admittances of the sub-conductors. This value is obtaining using:

$$\frac{1}{Z_{EQ\,int}} = \sum_{i=1}^{n} \frac{1}{Z_{i-int}} \tag{38}$$

In case of equal sub-conductors, the last equation is simplified and it is based on the internal impedance of any sub-conductor of the bundle ($Z_{SUB\,int}$). It is obtained from the next equation:

$$Z_{EQ\,int} = \frac{Z_{SUB\,int}}{n} \tag{39}$$

The equivalent shunt admittance is equal to the shunt capacitance and it is calculated by:

$$C_{EQ} = 2\pi\varepsilon_0 \frac{1}{\ln\left(\dfrac{2h}{r_{GMR}}\right)} \Rightarrow Y_{EQ} = j\omega \frac{2\pi\varepsilon_0}{\ln\left(\dfrac{2h}{r_{GMR}}\right)} \tag{40}$$

II.3. ALTERNATIVE METHOD FOR DETERMINATION OF EQUIVALENT CONDUCTOR FROM A BUNDLED CONDUCTOR [36]

The generic bundled conductor shown in Figure II.2 is shown again in the next figure, considering the line sending and receiving terminals as well as a non-ideal ground. For sub-conductor 1, it is related the currents I_{A1} and I_{B1} that are associated to the terminal A and terminal B, respectively. These relations are similar to the other sub-conductors. The I_A current is the sum of all sub-conductor currents in terminal A and the I_B current is the sum of all sub-conductor currents in terminal B. It is shown in the next equation. In Figure II.3, for the all sub-conductors, the voltage level is V_A in terminal A and V_B in terminal B. Basing on this and using distributed line parameters, it is obtained the relationships among voltages and currents shown in the next equations. These equations can be related to an equivalent conductor that represents the bundled conductor.

$$V_A = V_B \cosh(\gamma \cdot d_L) - Z_C \cdot I_B \sinh(\gamma \cdot d_L)$$

$$I_A = \frac{V_B}{Z_C} \cosh(\gamma \cdot d_L) - I_B \sinh(\gamma \cdot d_L)$$

(41)

In the last equations, d_L is the length of the equivalent conductor that is equal to the length of each sub-conductor. The propagation function (γ) and the characteristic impedance (Z_C) are calculated using the equivalent longitudinal impedance and the equivalent shunt admittance.

$$\gamma = \sqrt{Z_{EQ} \cdot Y_{EQ}} \quad and \quad Z_C = \sqrt{\frac{Z_{EQ}}{Y_{EQ}}}$$

(42)

The longitudinal impedance and the shunt admittance matrices, Z_{SUB} and Y_{SUB}, respectively, which represent the characteristics of sub-conductors, are frequency dependent. In this case, there are couplings among the sub-conductors. Because of this, the elements that are out of the mentioned matrices' main diagonal are not null. The elements of Z_{SUB} and Y_{SUB} are calculated following the development shown in item II.1. If it is set up the relationships between these matrices and the equivalent conductor characteristics, it is possible to carry out the procedure proposed in this item. It is made using the modal transformation applied to the Z_{SUB} and Y_{SUB} matrices. Manipulating the obtained mode matrices, it is set up the mentioned relationships though the Z_C and γ values. In this case, the modal transformation is used because there are not mutual couplings in mode domain. In this domain, a bundled conductor with n sub-conductors is represented by n uncoupled propagation modes.

$$Z_{SUB} = \begin{bmatrix} Z_{11} & \cdots & Z_{1n} \\ \vdots & \ddots & \vdots \\ Z_{1n} & \cdots & Z_{nn} \end{bmatrix} \quad and \quad Y_{SUB} = \begin{bmatrix} Y_{11} & \cdots & Y_{1n} \\ \vdots & \ddots & \vdots \\ Y_{1n} & \cdots & Y_{nn} \end{bmatrix} \tag{43}$$

The mode matrices are:

$$Z_{SUB\,M} = \begin{bmatrix} Z_{M1} & 0 & \cdots & 0 \\ 0 & Z_{M2} & \vdots & \vdots \\ \vdots & \vdots & \ddots & 0 \\ 0 & \cdots & 0 & Z_{Mn} \end{bmatrix} = T^T \cdot Z \cdot T$$

$$Y_{SUB\,M} = \begin{bmatrix} Y_{M1} & 0 & \cdots & 0 \\ 0 & Y_{M2} & \vdots & \vdots \\ \vdots & \vdots & \ddots & 0 \\ 0 & \cdots & 0 & Y_{Mn} \end{bmatrix} = T^{-1} \cdot Y \cdot T^{-T} \tag{44}$$

The $Z_{SUB\,M}$ matrix is the mode longitudinal impedance one. The $Y_{SUB\,M}$ matrix is the shunt admittance one. The T matrix is the eigenvector matrix obtained from the YZ matricial product. Using the mode domain, the propagation modes are uncoupled and can be independently modeled applying the next equations where it is considered a generic propagation mode related to the analyzed bundled conductor:

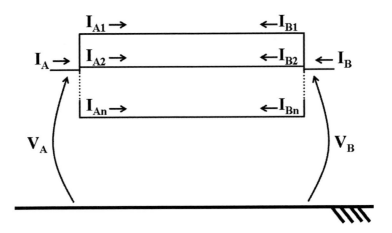

Figure II.3. Sending and receiving terminals of bundled conductor.

$$
\left.\begin{aligned}
V_{MAi} &= V_{MBi}\cosh\!\left(\gamma_{Mi}\cdot d_L\right) - Z_{CMi}\cdot I_{MBi}\sinh\!\left(\gamma_{Mi}\cdot d_L\right) \\[2mm]
I_{MAi} &= \frac{V_{MBi}}{Z_{CMi}}\cosh\!\left(\gamma_{Mi}\cdot d_L\right) - I_{MBi}\sinh\!\left(\gamma_{Mi}\cdot d_L\right)
\end{aligned}\right\},\quad i = 1,\,...,\,n
$$

$$(45)$$

In this case, Z_{CMi} and γ_{Mi} are the characteristic impedance and the propagation function of a generic propagation mode (i mode). The currents associated to this mode are I_{MAi} and I_{MBi}. The voltages are V_{MAi} and V_{MBi}. From the mode matrices, it is written:

$$
\gamma_{Mi} = \sqrt{Z_{Mi}\cdot Y_{Mi}} \quad and \quad Z_{CMi} = \sqrt{\frac{Z_{Mi}}{Y_{Mi}}}
\tag{46}
$$

Using matricial equations, it can be obtained the following linear system:

$$
V_{MA} = \theta_1\cdot V_{MB} - \theta_2\cdot I_{MB} \quad and \quad I_{MA} = \theta_3\cdot V_{MB} - \theta_4\cdot I_{MB}
\tag{47}
$$

Basing on the equations (45) and (47), it is defined:

$$
V_{MA} = \begin{bmatrix} V_{MA1} \\ \vdots \\ V_{MAn} \end{bmatrix},\;
I_{MA} = \begin{bmatrix} I_{MA1} \\ \vdots \\ I_{MAn} \end{bmatrix},\;
V_{MB} = \begin{bmatrix} V_{MB1} \\ \vdots \\ V_{MBn} \end{bmatrix},\;
I_{MB} = \begin{bmatrix} I_{MB1} \\ \vdots \\ I_{MBn} \end{bmatrix}
\tag{48}
$$

It is also defined:

$$
\theta_1(i,i) = \cosh\!\left(\gamma_{Mi}\cdot d_L\right)
$$

$$
\theta_2(i,i) = Z_{CMi}\sinh\!\left(\gamma_{Mi}\cdot d_L\right)
$$

$$(48)$$

$$
\theta_3(i,i) = \frac{\cosh\!\left(\gamma_{Mi}\cdot d_L\right)}{Z_{CMi}} = \frac{\theta_1(i,i)}{Z_{CMi}}
$$

$$
\theta_4(i,i) = \sinh\!\left(\gamma_{Mi}\cdot d_L\right) = \frac{\theta_2(i,i)}{Z_{CMi}}
$$

The θ_1, θ_2, θ_3 and θ_4 matrices are diagonal matrices where only the main diagonal elements are not null.

The vectors shown in equations (48) are related to the phase domain using the following:

$$V_{MA} = T^T \cdot V_A , \quad I_{MA} = T^{-1} \cdot I_A$$

$$V_{MB} = T^T \cdot V_B , \quad I_{MB} = T^{-1} \cdot I_B$$

(49)

Using these relations, the voltages and currents are converted to the phase domain. If it is applied to the obtained linear system, it is obtained:

$$V_A = T^{-T} \cdot \theta_1 \cdot T^T \cdot V_B - T^{-T} \cdot \theta_2 \cdot T^{-1} \cdot I_B$$

$$I_A = T \cdot \theta_3 \cdot T^T \cdot V_B - T \cdot \theta_4 \cdot T^{-1} \cdot I_B$$

(50)

Manipulating the last equations, it is obtained:

$$I_B = T \cdot \theta_2^{-1} \cdot \theta_1 \cdot T^T \cdot V_B - T \cdot \theta_2^{-1} \cdot T^T \cdot V_A$$
$$\Downarrow$$
$$I_A = T \cdot \theta_4 \cdot \theta_2^{-1} \cdot T^T \cdot V_A + \left(T \cdot \theta_3 \cdot T^T - T \cdot \theta_4 \cdot \theta_2^{-1} \cdot \theta_1 \cdot T^T \right) \cdot V_B$$

(51)

Simplifying the equation related to the I_A current matrix, it is obtained:

$$I_A = A \cdot V_A + B \cdot V_B$$

(52)

In this case, A and B matrices are defined as:

$$A = T \cdot \theta_4 \cdot \theta_2^{-1} \cdot T^T \quad and \quad B = T \cdot \theta_3 \cdot T^T - T \cdot \theta_4 \cdot \theta_2^{-1} \cdot \theta_1 \cdot T^T$$

(53)

Describing the last equations in detail, it is obtained:

$$\begin{bmatrix} I_{A1} \\ \vdots \\ I_{An} \end{bmatrix} = \begin{bmatrix} A_{11} & \cdots & A_{1n} \\ \vdots & \ddots & \vdots \\ A_{n1} & \cdots & A_{nn} \end{bmatrix} \cdot \begin{bmatrix} V_A \\ \vdots \\ V_A \end{bmatrix} + \begin{bmatrix} B_{11} & \cdots & B_{1n} \\ \vdots & \ddots & \vdots \\ B_{n1} & \cdots & B_{nn} \end{bmatrix} \cdot \begin{bmatrix} V_B \\ \vdots \\ V_B \end{bmatrix}$$

(54)

The addition of the linear system equations shown in equation (54) leads to:

$$I_A = \sum_{i=1}^{n}\sum_{k=1}^{n} A_{ik} \cdot V_A + \sum_{i=1}^{n}\sum_{k=1}^{n} B_{ik} \cdot V_B \tag{55}$$

The linear system that describes the bundled conductor is rewritten as following:

$$V_B = V_A \cosh(\gamma \cdot d_L) - Z_C \cdot I_A \sinh(\gamma \cdot d_L)$$

$$I_B = \frac{V_A}{Z_C}\cosh(\gamma \cdot d_L) - I_A \sinh(\gamma \cdot d_L) \tag{57}$$

Manipulating the first equation of the shown linear system, it is obtained the I_A current as a function of only the V_A and V_B voltages:

$$I_A = \frac{\cosh(\gamma \cdot d_L)}{Z_C \cdot \sinh(\gamma \cdot d_L)} \cdot V_A - \frac{1}{Z_C \cdot \sinh(\gamma \cdot d_L)} \cdot V_B \tag{58}$$

Equaling the equations (55) and (58), it is determined the propagation function and the characteristic impedance from the elements of the A and B matrices as following:

$$\gamma = \frac{1}{d_L} \operatorname{arc\,cosh}\left(-\frac{\sum_{i=1}^{n}\sum_{k=1}^{n} A_{ik}}{\sum_{i=1}^{n}\sum_{k=1}^{n} B_{ik}} \right) \tag{59}$$

$$Z_C = \frac{1}{\sqrt{\left(\sum_{i=1}^{n}\sum_{k=1}^{n} A_{ik}\right)^2 - \left(\sum_{i=1}^{n}\sum_{k=1}^{n} B_{ik}\right)^2}}$$

These values can be applied to the determination of the equivalent conductor characteristics. It is shown in the next equations.

$$Z_{EQ} = \gamma \cdot Z_C = \frac{1}{d_L \cdot \sqrt{\left(\sum_{i=1}^{n}\sum_{k=1}^{n} A_{ik}\right)^2 - \left(\sum_{i=1}^{n}\sum_{k=1}^{n} B_{ik}\right)^2}} \operatorname{arc cosh}\left(-\frac{\sum_{i=1}^{n}\sum_{k=1}^{n} A_{ik}}{\sum_{i=1}^{n}\sum_{k=1}^{n} B_{ik}}\right) \quad (60)$$

$$Y_{EQ} = \frac{\gamma}{Z_C} = \frac{\sqrt{\left(\sum_{i=1}^{n}\sum_{k=1}^{n} A_{ik}\right)^2 - \left(\sum_{i=1}^{n}\sum_{k=1}^{n} B_{ik}\right)^2}}{d_L} \operatorname{arc cosh}\left(-\frac{\sum_{i=1}^{n}\sum_{k=1}^{n} A_{ik}}{\sum_{i=1}^{n}\sum_{k=1}^{n} B_{ik}}\right)$$

Simplifying the terms, it is defined one term for the addition of all the A matrix elements and other term for the addition of all B matrix elements:

$$S_A = \sum_{i=1}^{n}\sum_{k=1}^{n} A_{ik} \quad and \quad S_B = \sum_{i=1}^{n}\sum_{k=1}^{n} B_{ik} \quad (61)$$

Applying these terms, the equations (58) are rewritten. The Z_{EQ} value is:

$$Z_{EQ} = \frac{1}{d_L \cdot \sqrt{S_A^2 - S_B^2}} \operatorname{arc cosh}\left(-\frac{S_A}{S_B}\right) \quad (62)$$

The Y_{EQ} value is:

$$Y_{EQ} = \frac{\sqrt{S_A^2 - S_B^2}}{d_L} \operatorname{arc cosh}\left(-\frac{S_A}{S_B}\right) \quad (63)$$

From these obtained equivalent values, it is determined the conductor characteristics that is equivalent to a bundled conductor. Using the Z_{EQ} and Y_{EQ} values, the R_{EQ}, L_{EQ} and C_{EQ} values can be obtained for a determined frequency range, defining the equivalent conductor for the analyzed bundled conductor.

As a sample for applying the proposed procedure, it is shown in the next figure a non-conventional bundled conductor that is composed by 7 sub-conductors. There is a central conductor with a 3.5 cm radius. The radius of each surrounding sub-conductor is 1.5 cm. The ground resistivity is considered as 1000 Ω·m. The height related to the geometrical center of this bundled conductor is 12 m and it corresponds to the height of the central sub-conductor.

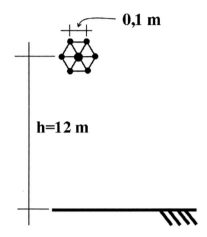

Figure II.4. Non-conventional bundled conductor.

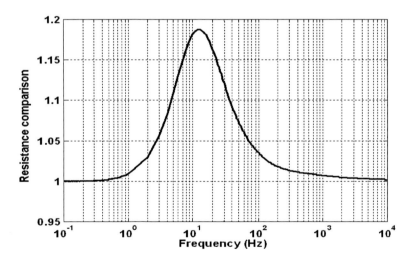

Figure II.5. Comparisons of resistance values calculated from the GMR method and the proposed procedure for the equivalent conductor.

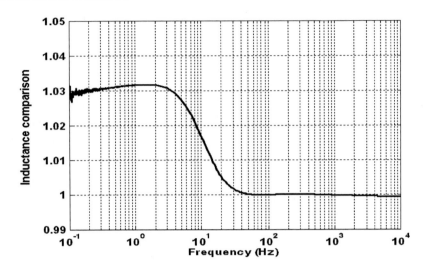

Figure II.6. Comparisons of inductance values calculated from the GMR method and the proposed procedure for the equivalent conductor.

The parameters of the equivalent conductor that represents the shown bundled conductor are calculated using the GMR method and the procedure proposed in this chapter. After this, the results are compared and shown in the previous both figures. In this case, Figure II.5 show the comparisons obtained for resistance values. In the frequency range from 10 Hz to 100 Hz, the difference between the two used methods reaches the highest value that is about 18%. Then, the unbalanced distribution of currents leads to a great difference in determination of the resistance value for the steady state frequency range. It is also concluded for the comparisons results of inductance values shown in Figure II.6. From low frequencies to 5 Hz, there is approximately a constant difference between the both methods. This difference is about 3%.

For bundled conductors with balanced current distribution, the proposed procedure and the GMR method get equal results. Basin on the results of this chapter, the GMR method is not adequate, if there is unbalanced current distribution on the analyzed bundle conductor. Then, the proposed procedure is a useful numeric tool, because it is adequate for calculating the equivalent conductor of bundled conductors when these have balanced and unbalanced current distribution.

Chapter 3

UNTRANSPOSED SYMMETRICAL THREE-PHASE TRANSMISSION LINE MODAL REPRESENTATION USING TWO TRANSFORMATION MATRICES [37, 38]

It is proposed in this chapter a modal representation for untransposed symmetrical three-phase transmission lines basing on the application of two transformation matrices. In this case, it is considered transmission lines that have a vertical symmetry plane. The first applied transformation matrix is Clarke's matrix. This matrix decomposes the line phases in one exact mode and two non-exact modes, called quasi-modes. From these quasi-modes, it is obtained the other two exact modes applying a 2-order modal transformation matrix. The direct application of modal transformation for three-phase transmission lines uses 3-order frequency dependent transformation matrices. So, the advantage of the proposed way for modal transformation application is the reduction of the frequency dependent elements that compose this transformation. In this case, Clarke's matrix is frequency independent and it is necessary only the 2-order modal transformation matrix application.

For electromagnetic transient analyses in power systems, it can be considered numeric methods based on the phase domain or modal domain. Applying the modal domain, usually, it is considered a hybrid model where it is included the interactions between the phase domain and the mode domain using the modal transformations. The wave propagation on the line is determined in mode domain and the interactions with the phase domain are necessary for including the obtained results in the remaining network power system.

When there is a vertical symmetry plane in three-phase transmission lines and these lines are not transposed, the Clarke's matrix application obtains the α, β and 0 components. The β component is an exact mode because it does not have couplings with the other both components. Between the other both components, α and 0, there is a coupling and they are not be considered exact modes. Because of this, some authors have called these components as the α and 0 quasi-modes. For some cases, this coupling can be neglected and the α and 0 exact modes can be changed into the α and 0 quasi-modes, respectively. For general symmetrical three-phase transmission line cases, the α and 0 quasi-modes can be treated as a two-phase transmission line and this hypothetical line is changed into two uncoupled two exact modes.

III.1. Obtaining the β Exact Mode Applying Clarke's Matrix

Considering an unstransposed three-phase transmission line with a vertical symmetry plane, it is used the next figure schematizing this situation. The central phase height is different of the height of adjacent phases. So, the heights of adjacent phases are equal. The horizontal distance between each adjacent phase and the central phase, d_{HD} in the next figure, is also equal.

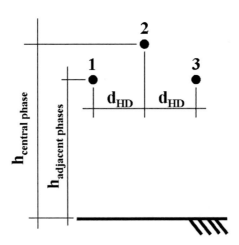

Figure III.1. A symmetrical three-phase transmission line schema.

For the transmission line schematize in the last figure, the Z and Y matrices can be described as following:

$$Z_S = \begin{bmatrix} Z_{11} & Z_{12} & Z_{13} \\ Z_{12} & Z_{22} & Z_{12} \\ Z_{13} & Z_{12} & Z_{11} \end{bmatrix} \quad and \quad Y_s = \begin{bmatrix} Y_{11} & Y_{12} & Y_{13} \\ Y_{12} & Y_{22} & Y_{12} \\ Y_{13} & Y_{12} & Y_{11} \end{bmatrix} \tag{64}$$

Because the symmetry plane, the characteristics related to the phase 1 of Figure III.1 are equal to those related to the phase 3 of the same figure. It leads to the results shown in the last equation. Applying Clarke's matrix to the Z_S and Y_S matrices, it is obtained these matrices in the new domain. It is shown in equations (65). This transformation is not the exact modal transformation because Clarke's matrix is not an eigenvector matrix for the considered case. Completing the modal transformation, it is only considered the α and 0 quasi-modes. It is used this denomination because there is coupling between these two components.

$$Z_{MS} = T_{CL}^T \cdot Z_{SL} \cdot T_{CL} = \begin{bmatrix} Z_{\varrho\alpha} & 0 & Z_{\alpha 0} \\ 0 & Z_\beta & 0 \\ Z_{\alpha 0} & 0 & Z_{\varrho 0} \end{bmatrix}$$

$$Y_{MS} = T_{CL}^T \cdot Y_{SL} \cdot T_T = \begin{bmatrix} Y_{\varrho\alpha} & 0 & Y_{\alpha 0} \\ 0 & Y_\beta & 0 \\ Y_{\alpha 0} & 0 & Y_{\varrho 0} \end{bmatrix} \tag{65}$$

The structure of Clarke's matrix is:

$$T_{CL} = \begin{bmatrix} -\dfrac{1}{\sqrt{6}} & \dfrac{1}{\sqrt{2}} & \dfrac{1}{\sqrt{3}} \\ \dfrac{2}{\sqrt{6}} & 0 & \dfrac{1}{\sqrt{3}} \\ -\dfrac{1}{\sqrt{6}} & -\dfrac{1}{\sqrt{2}} & \dfrac{1}{\sqrt{3}} \end{bmatrix} \tag{66}$$

The β component obtained from the Clarke's matrix application is an exact mode because the $Z_{\alpha\beta}$, $Z_{\beta 0}$, $Y_{\alpha\beta}$, $Y_{\beta 0}$ couplings in equations (65) are null. Separating the coupled quasi-modes, it is obtained:

$$Z_{Q\alpha0} = \begin{bmatrix} Z_{Q\alpha} & Z_{\alpha0} \\ Z_{\alpha0} & Z_{Q0} \end{bmatrix} \quad and \quad Y_{Q\alpha0} = \begin{bmatrix} Y_{Q\alpha} & Y_{\alpha0} \\ Y_{\alpha0} & Y_{Q0} \end{bmatrix} \tag{67}$$

III.2. OBTAINING THE α AND 0 EXACT MODES APPLYING A 2-ORDER TRANSFORMATION MATRIX

The $Z_{Q\alpha0}$ and $Y_{Q\alpha0}$ can be represented as a hypothetic two-phase transmission line that has not a symmetry plane. This hypothetic line is shown in the next figure and the angle between the both phases ($\theta_{\alpha0}$) should be different of 0 or 180 °. In case of these values, the couplings between α and 0 quasi-modes are null.

From the $Z_{Q\alpha0}$ and $Y_{Q\alpha0}$ matrices, it is obtaining the eigenvector matrix of the $Y_{Q\alpha0}Z_{Q\alpha0}$ matricial product identified as the $T_{\alpha0}$ transformation matrix. Applying this matrix, it is obtaining the exact α and 0 modes as showing in the following:

$$Z_{M\alpha0} = T_{\alpha0}^{-1} \cdot Z_{Q\alpha0} \cdot T_{\alpha0} = \begin{bmatrix} Z_\alpha & 0 \\ 0 & Z_0 \end{bmatrix}$$

$$Y_{M\alpha0} = T_{\alpha0}^{-1} \cdot Y_{Q\alpha0} \cdot T_{\alpha0} = \begin{bmatrix} Y_\alpha & 0 \\ 0 & Y_0 \end{bmatrix} \tag{68}$$

The $T_{\alpha0}$ matrix structure is:

$$T_{\alpha0} = \begin{bmatrix} T_{\alpha0\,11} & T_{\alpha0\,21} \\ T_{\alpha0\,12} & T_{\alpha0\,22} \end{bmatrix} \tag{69}$$

Recomposing the three-order transformation matrix, it is considered the $T_{\alpha0}$ matrix elements as following:

$$T_{3\alpha0} = \begin{bmatrix} T_{\alpha0\,11} & 0 & T_{\alpha0\,21} \\ 0 & 1 & 0 \\ T_{\alpha0\,12} & 0 & T_{\alpha0\,22} \end{bmatrix} \tag{70}$$

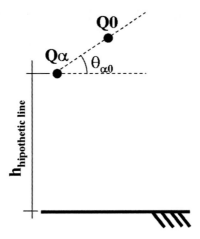

Figure III.2. Hypothetic line for the coupled α and 0 quasi-mode components.

The modal transformation matrix for the symmetrical three-phase transmission line considering the untransposed case is obtaining as:

$$T_{SL} = T_{CL} \cdot T_{3\alpha 0} = \begin{bmatrix} \left(\dfrac{-T_{\alpha 0\,11}}{\sqrt{6}} + \dfrac{T_{\alpha 0\,12}}{\sqrt{3}} \right) & \dfrac{1}{\sqrt{2}} & \left(\dfrac{-T_{\alpha 0\,21}}{\sqrt{6}} + \dfrac{T_{\alpha 0\,22}}{\sqrt{3}} \right) \\ \left(\dfrac{2 \cdot T_{\alpha 0\,11}}{\sqrt{6}} + \dfrac{T_{\alpha 0\,12}}{\sqrt{3}} \right) & 0 & \left(\dfrac{2 \cdot T_{\alpha 0\,11}}{\sqrt{6}} + \dfrac{T_{\alpha 0\,12}}{\sqrt{3}} \right) \\ \left(\dfrac{-T_{\alpha 0\,11}}{\sqrt{6}} + \dfrac{T_{\alpha 0\,12}}{\sqrt{3}} \right) & \dfrac{-1}{\sqrt{2}} & \left(\dfrac{-T_{\alpha 0\,21}}{\sqrt{6}} + \dfrac{T_{\alpha 0\,22}}{\sqrt{3}} \right) \end{bmatrix} \quad (71)$$

For symmetrical three-phase transmission lines, using the proposed procedure, the modal transformation matrix is obtained from the matricial product between Clarke's matrix and the $T_{3\alpha 0}$ one. This last is calculated based on the α and 0 quasi-modes. After the Clarke's matrix application, these quasi-modes have coupled yet and it is necessary manipulated then for obtaining the exact α and 0 modes. For this manipulation, it is obtained the elements of the $T_{3\alpha 0}$ matrix that completes the modal transformation for the analyzed case.

The advantage of the proposed procedure is the determination of only four frequency dependent elements that composed the modal transformation matrix. If it is used a procedure adequate to a general three-phase transmission line case, it is necessary all nine elements of the modal transformation matrix that depends on the frequency.

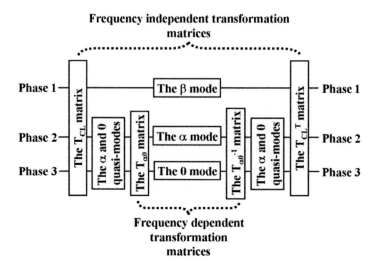

Figure III.3. Schema for applying the procedure proposed in this chapter.

The schema shown in Figure III.3 illustrates the procedure proposed in this item. Applying this schematized procedure for electromagnetic transient and wave propagation simulation can imply in convolution numeric method uses. The convolution numeric methods are necessary for solving the interactions between the phase domain and mode domain through the frequency transformation matrices. So, the proposed procedure can be an alternative numeric method for changing the phase domain values into mode domain values. Then, a actual three-phase line that belongs to Brazilian's utilities, is a sample for applying the proposed procedure and shown in Figure III.4.

III.3. AN ACTUAL TRANSMISSION LINE SAMPLE

About the the line shown in the next figure, it is a 440 kV transmission line with a vertical symmetry plane. The central phase conductor height is 27.67 m on the tower. The height of adjacent phase conductors is 24.07 m. Every phase is composed of four sub-conductors distributed in a square shape with 0.4 m side length. Every sub-conductor is an ACSR type one (ACSR 26/7 636 MCM) with internal diameter of 0.93 cm and a external diameter of the 2.52 cm. The phase sub-conductor resistivity is 0.089899 Ω/km and the sag at the midspan is 13.43 m. The earth resistivity is considered constant (1000 Ω.m). There are two ground wires and they are EHS 3/8" with the resistivity of 4.188042 Ω/km. The diameters of these cables are 0.9144 cm. The height of these cables on the tower is 36.00 m. The sag of the ground wires at the midspan is 6.40 m. Each ground wire is composed by a single conductor.

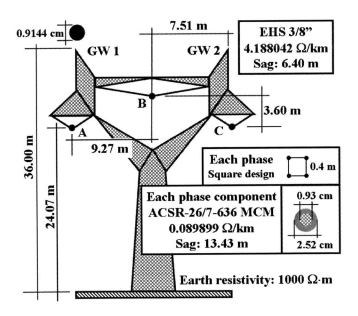

Figure III.4. An actual transmission line.

The next figure shows the real part of mode impedances calculated using the procedure proposed in this chapter.

In the next figure, it is shown the results obtaining for mode inductances that are the imaginary part of the mode impedances. In this case, it is also used the proposed procedure.

For checking the efficiency and the accuracy of the proposed procedure, it is carried out a frequency scan testing mode domain using a 1 pu voltage step source linked to the phase 1 of the shown actual transmission line. The sending terminals of the other both phases are in short-circuit. The receiving terminals of all phases are linked to the impedances which values are equal to the line characteristic impedance (Z_C). It avoids the propagated wave reflections. Applying the proposed procedure, it is calculated the modes of the actual transmission line sample and the 1 pu voltage source is inserted. This is schematized in Figure III.7.

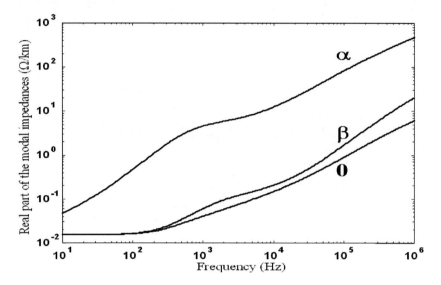

Figure III.5. Mode resistances of the actual transmission line used as a sample.

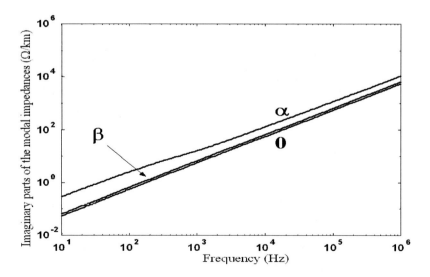

Figure III.6. Mode inductances of the actual transmission line used as a sample.

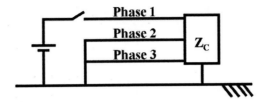

Figure III.7. Energization of the actual transmission line phase 1.

The obtained results of the mentioned test are compared to those obtained from the 3-order frequency dependent modal transformation matrix application. This matrix is calculated using Newton-Raphson's numeric method. Both obtained results are compared to those related to Clarke's matrix. The next figure shows these comparisons that are related to the voltage values that depend on the frequency for the phase 2 receiving terminal.

Based on the comparisons shown in Figure III.8, the proposed procedure can be classified an exact numeric method for calculating the modal transformation of a three-phase transmission line with a vertical symmetry plane. Besides this, it is observed that there are differences among the results related to Clarke's matrix and the exact results obtained from the both exact methods until 10 kHz. Discussing if these differences are significant or not, it is analyzed in the next chapters the application of Clarke's matrix for electromagnetic transient simulations considering untransposed transmission lines.

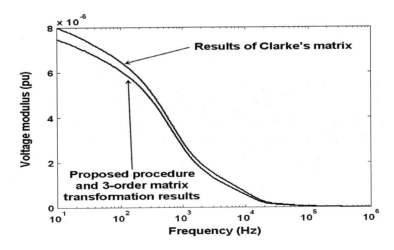

Figure III.8. Comparisons of frequency scan test results in mode domain.

Chapter 4

SINGLE REAL TRANSFORMATION MATRIX APPLICATIONS FOR UNTRANSPOSED THREE-PHASE TRANSMISSION LINES [39-41]

Analyzing transposed three-phase transmission lines, single real matrices can be carried out exact modal transformations because the mentioned matrices are eigenvector ones in these cases. For untransposed three-phase transmission lines, if single real matrices are used as transformation ones, it is not obtained exact modal transformations. However, single real transformation matrices can be good approximations for the exact transformation matrices and it is analyzed in this chapter considering three-phase lines.

Considering transmission line analyses, the frequency influence is significant for line longitudinal parameters. Because of this, the line representation in time domain becomes complex. There are models where it is applied phase mode or modal transformation, considering the modeling in mode domain and improving the frequency dependent line parameter representation. For this, it is necessary to take to frequency dependent transformation matrices, because the line characteristics depend on the frequency. Frequency independent transformation matrices can be an alternative for this analysis type, if they lead to good approximations for the exact transformations. The interest about this alternative is based on these characteristics of the mentioned constant matrix application: single, frequency independent, line parameter independent. The main explanation for the interest in the frequency independent matrix application as a modal transformation matrix is that, considering transposed transmission line, the modal transformation matrices are frequency independent and, the most of these cases, they are also line parameter independent.

There are three main matrices that have used as modal transformation matrices for three-phase transmission: Clarke's matrix, Fortescue's one and Karrenbauer's one. These matrices are exact eigenvector and modal transformation matrices for transposed three-phase transmission lines. For three-phase systems, Fortescue's matrix is degenerated into symmetrical component one. Clarke's matrix is a linear combination of the symmetrical component matrix, searching for real matrix elements. So, Clarke's matrix, besides the mentioned characteristics, incorporated more two characteristics into the application analyzed in this chapter: this matrix is real and its application is identical for voltage and current manipulating. With Clarke's matrix use, mathematical model simplifications are obtained and the developed model can be applied directly in time domain programs without numeric convolution methods procedures.

In this chapter, for typical untransposed three-phase transmission lines, the eigenvector and eigenvalue analyses are carried out using Clarke's matrix as an alternative for modal transformation matrix. It is considered symmetrical and asymmetrical three-phase transmission line cases. When the Clarke's matrix results, called quasi-modes, are compared to the eigenvalues, the relative errors are reasonably small and can be considered negligible as well as accepted for most applications.

IV.1. MATHEMATICAL MODELING FOR MODAL TRANSFORMATION APPLICATION

When transmission lines are submitted to electromagnetic transient phenomena, the relations among transversal voltage vector (V_{PH}) and longitudinal current one (I_{PH}) in phase domain can be described by the following equations. As previously mentioned, the Z and Y matrices are the longitudinal impedance and shunt admittance ones, respectively.

$$-\frac{dV_{PH}}{dx} = Z \cdot I_{PH} \quad and \quad -\frac{dI_{PH}}{dx} = Y \cdot V_{PH} \qquad (72)$$

In mode domain, the V_{PH} and I_{PH} vectors are changed into the mode voltage (V_M) and the mode current (I_M) vectors, respectively, using the following:

$$V_{PH} = T_V \cdot V_M \quad and \quad I_{PH} = T_I \cdot I_M \tag{73}$$

The T_I matrix is obtained from equation (2). The T_V matrix is obtained from the next equation.

$$T_V^{-1} \cdot Z \cdot Y \cdot T_V = \lambda \tag{74}$$

Equaling and manipulating equations (2) and (74), it is obtained:

$$T_V^{-1} = T_I^T \tag{75}$$

Applying these results on equations (72), it is obtained:

$$-\frac{d(T_V \cdot V_M)}{dx} = Z \cdot T_I \cdot I_M \quad and \quad -\frac{d(T_I \cdot I_M)}{dx} = Y \cdot T_V \cdot V_M$$

$$\Downarrow \tag{76}$$

$$-\frac{d V_M}{dx} = T_I^T \cdot Z \cdot T_I \cdot I_M \quad and \quad -\frac{d I_M}{dx} = T_I^{-1} \cdot Y \cdot T_I^{-T} \cdot V_M$$

And these last equations also lead to the relations shown in equations (4). In this item, for three-phase transmission line cases, equations (4) are modified changing the exact eigenvector matrices into Clarke's matrix as described following:

$$T_V = T_{CL} \quad and \quad T_I = T_{CL} \tag{77}$$

Clarke's inverse matrix is Clarke's transposed matrix:

$$T_{CL}^{-1} = T_{CL}^T \tag{78}$$

With the mentioned changes, it is obtained:

$$Z_{MCL} = T_{CL}^T \cdot Z \cdot T_{CL} \quad and \quad Y_{MCL} = T_{CL}^T \cdot Y \cdot T_{CL} \tag{79}$$

It is also obtained:

$$\lambda_{ICL} = T_{CL}^T \cdot Y \cdot Z \cdot T_{CL} \quad and \quad \lambda_{VCL} = T_{CL}^T \cdot Z \cdot Y \cdot T_{CL} \tag{80}$$

For general three-phase transmission line cases, the λ_{ICL} and λ_{VCL} matrices are not equal because the ZY and YZ matricial products are neither equal. Changing the eigenvector matrices into Clarke's matrix, the modal transformation is not exact, the quasi-modes are obtained and the previously mentioned matrices, Z_{MCL}, Y_{MCL}, λ_{ICL} as well as λ_{VCL}, are not diagonal ones.

For transposed three-transmission line cases, considering ideal assumptions, the YZ and ZY matricial products become equal, becoming equal the Z_{MCL} and Y_{MCL} matrices to the exact correspondent ones as well as the λ_{ICL} and λ_{VCL} matrices to the λ matrix. In this case, it is obtained the following equations where the T letter addition in subscript term identifies the transposed cases:

$$Z_{TM} = T_{CL}^T \cdot Z \cdot T_{CL} = Z_M = T_I^T \cdot Z \cdot T_I = \begin{bmatrix} Z_\alpha & 0 & 0 \\ 0 & Z_\beta & 0 \\ 0 & 0 & Z_0 \end{bmatrix} \tag{81}$$

$$Y_{TM} = T_{CL}^T \cdot Y \cdot T_{CL} = Z_M = T_I^T \cdot Y \cdot T_I = \begin{bmatrix} Y_\alpha & 0 & 0 \\ 0 & Y_\beta & 0 \\ 0 & 0 & Y_0 \end{bmatrix}$$

For the λ_{ICL} and λ_{VCL} matrices, the equations are changed into:

$$\lambda_{TI} = T_{CL}^T \cdot Y \cdot Z \cdot T_{CL} = \lambda_{TV} = T_{CL}^T \cdot Z \cdot Y \cdot T_{CL} = \lambda = \begin{bmatrix} \lambda_\alpha & 0 & 0 \\ 0 & \lambda_\beta & 0 \\ 0 & 0 & \lambda_0 \end{bmatrix} \tag{82}$$

In sequence, it is analyzed the application of Clarke's matrix to typical untransposed three-phase transmission line cases.

IV.2. UNTRANSPOSED THREE-PHASE LINES WITH A VERTICAL SYMMETRY PLANE

This transmission line type can be represented by schema shown in Figure III.1. Considering the couplings among the phases and the vertical symmetrical plane, the longitudinal impedance and shunt admittance matrices in phase domain are described by equations (64). Applying Clarke's matrix, the same matrices in mode domain are described by equations (65). The quasi-mode matrices are described as following:

$$\lambda_{IS} = \begin{bmatrix} \lambda_{IS\alpha} & 0 & \lambda_{IS\alpha0} \\ 0 & \lambda_{\beta} & 0 \\ \lambda_{IS\alpha0} & 0 & \lambda_{IS0} \end{bmatrix} \quad and \quad \lambda_{VS} = \begin{bmatrix} \lambda_{VS\alpha} & 0 & \lambda_{VS\alpha0} \\ 0 & \lambda_{\beta} & 0 \\ \lambda_{VS\alpha0} & 0 & \lambda_{VS0} \end{bmatrix} \tag{83}$$

The structure of the λ matrix is:

$$\lambda = T_I^{-1} \cdot Y \cdot Z \cdot T_I = T_I^T \cdot Z \cdot Y \cdot T_I^{-T} = \begin{bmatrix} \lambda_{\alpha} & 0 & 0 \\ 0 & \lambda_{\beta} & 0 \\ 0 & 0 & \lambda_0 \end{bmatrix} \tag{84}$$

Comparing the λ_{IS} and λ_{VS} matrix main diagonal elements to the correspondent λ matrix elements, it is used relative errors that are calculated by:

$$\varepsilon_K(\%) = \frac{\left(\lambda_{K\,QUASI-MODE} - \lambda_K\right)}{\lambda_K} \cdot 100, \quad K = \alpha, \beta \text{ and } 0 \tag{85}$$

In this case, the $\lambda_{K\,QUASI-MODE}$ values represent the main diagonal elements of the λ_{IS} matrix or the λ_{VS} one. The λ_K values are the non null elements of the λ matrix. The obtained relative errors are shown in the next both figures.

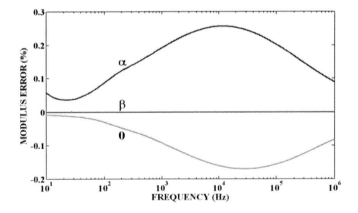

Figure IV.1. Comparisons among λ_{IS} quasi-modes and λ eigenvalues for the actual symmetrical three-phase line.

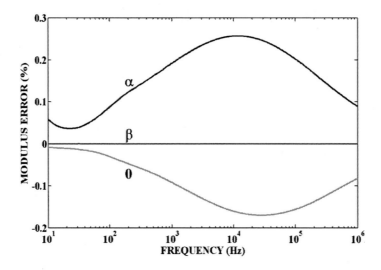

Figure IV.2. Comparisons among λ_{VS} quasi-modes and λ eigenvalues for the actual symmetrical three-phase line.

Based on the relative errors shown in the last both figures, it concluded that the λ_{IS} quasi-modes are equal to the λ_{VS} ones. It can also be considered that both λ_{IS} and λ_{VS} quasi-modes are practically equal to the eigenvalues, because the relative errors can be considered negligible.

The next both figures show results related to the Z and Y matrices.

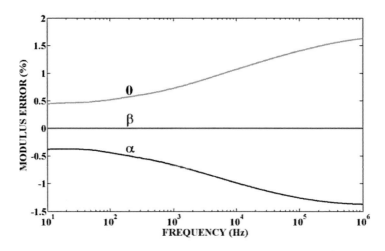

Figure IV.3. Relative errors related to the mode longitudinal impedances for the actual symmetrical three-phase transmission line.

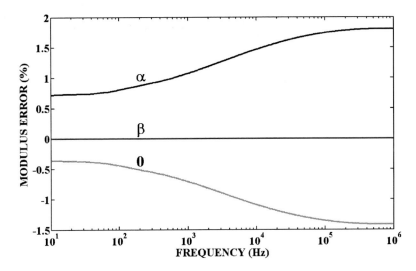

Figure IV.4. Relative errors related to the mode longitudinal admittances for the actual symmetrical three-phase transmission line.

Comparing the errors shown in last both figures to those related to the eigenvalues, these errors are about 10 times higher than the quasi-mode errors. For the relative error range, these values could be considered good approximations for the exact values yet. So, in this error range, equations (65) could be rewritten as:

$$Z_{MS} = T_{CL}^T \cdot Z_{SL} \cdot T_{CL} = \begin{bmatrix} Z_\alpha & 0 & Z_{\alpha 0} \\ 0 & Z_\beta & 0 \\ Z_{\alpha 0} & 0 & Z_0 \end{bmatrix}$$

(86)

$$Y_{MS} = T_{CL}^T \cdot Y_{SL} \cdot T_T = \begin{bmatrix} Y_\alpha & 0 & Y_{\alpha 0} \\ 0 & Y_\beta & 0 \\ Y_{\alpha 0} & 0 & Y_0 \end{bmatrix}$$

Equations (83) are rewritten as:

$$\lambda_{IS} = \begin{bmatrix} \lambda_\alpha & 0 & \lambda_{IS\alpha 0} \\ 0 & \lambda_\beta & 0 \\ \lambda_{IS\alpha 0} & 0 & \lambda_0 \end{bmatrix} \quad and \quad \lambda_{VS} = \begin{bmatrix} \lambda_\alpha & 0 & \lambda_{VS\alpha 0} \\ 0 & \lambda_\beta & 0 \\ \lambda_{VS\alpha 0} & 0 & \lambda_0 \end{bmatrix}$$

(87)

Completing the analyses carried out in this item, it is calculated the relative values of the off-diagonal elements of the λ_{IS} and λ_{VS} quasi-mode matrices. For this, it is used the next equations which results are shown in the next two figures.

$$\varepsilon_{I\alpha 0}(\%) = \frac{\lambda_{I\alpha 0}}{\lambda_\alpha \ or \ \lambda_0} \cdot 100 \quad and \quad \varepsilon_{V\alpha 0}(\%) = \frac{\lambda_{V\alpha 0}}{\lambda_\alpha \ or \ \lambda_0} \cdot 100 \qquad (88)$$

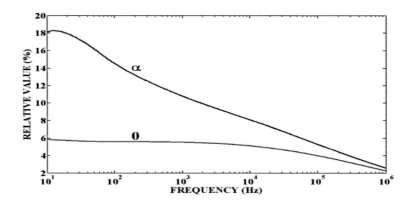

Figure IV.5. The $\lambda_{I\alpha 0}$ relative values for the actual symmetrical three-phase transmission line.

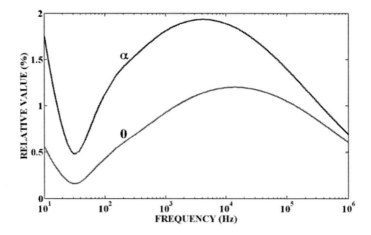

Figure IV.6. The $\lambda_{V\alpha 0}$ relative values for the actual symmetrical three-phase transmission line.

In Figure IV.5, it is shown the comparisons between the $\lambda_{I\alpha0}$ element and the λ eigenvalues that correspond to the λ_I coupled quasi-modes. Considering the $\lambda_{V\alpha0}$ element, the comparisons to the correspondent λ_V eigenvalues are shown in Figure IV.6. These results are lower than those of Figure IV.5. So, it is concluded that the T_{CL} matrix is a good approximation for the T_V matrix, considering the actual symmetrical three-phase transmission line. It is also concluded that the main problems related to the change of the eigenvector matrices into Clarke's matrix are related to the T_I eigenvector change, because the $\lambda_{I\alpha0}$ element reaches about 18 % of the λ_α eigenvalue. The influence of this coupling on electromagnetic transient analyses and simulations should be investigated. It is going carried out in the next specific item in which it is analyzed an asymmetrical three-phase line with triangular phase distribution.

Detailing these analyzes, the high relative value of the $\lambda_{I\alpha0}$ element can be lead to high errors, if this change is applied to calculating values such as the propagation function (γ) and characteristic impedance (Z_C). For the mentioned application, relative values can reach about 130 % when compared to the coupled quasi-mode values.

Because of these results, it is also investigated the application of a correction procedure for Clarke's matrix. Before this correction procedure, the results of the application of Clarke's matrix to the untransposed asymmetrical three-phase transmission lines are presented. It is shown two typical asymmetrical three-phase transmission line types.

IV.2. UNTRANSPOSED THREE-PHASE LINES WITH PHASE CONDUCTORS VERTICALLY LINED

Considering the line in the next figure and applying Clarke's matrix as modal transformation matrix, there are couplings between all three quasi-modes that compose the λ_{IR} and λ_{VR} quasi-mode matrices.

The mentioned matrices are described as:

$$\lambda_{IR} = \begin{bmatrix} \lambda_{IR\alpha} & \lambda_{IR\alpha\beta} & \lambda_{IR\alpha0} \\ \lambda_{IR\alpha\beta} & \lambda_{IR\beta} & \lambda_{IR\beta0} \\ \lambda_{IR\alpha0} & \lambda_{IR\beta0} & \lambda_{IR0} \end{bmatrix} \quad and \quad \lambda_{VR} = \begin{bmatrix} \lambda_{VR\alpha} & \lambda_{VR\alpha\beta} & \lambda_{VR\alpha0} \\ \lambda_{VR\alpha\beta} & \lambda_{VR\beta} & \lambda_{VR\beta0} \\ \lambda_{VR\alpha0} & \lambda_{VR\beta0} & \lambda_{VR0} \end{bmatrix} \quad (89)$$

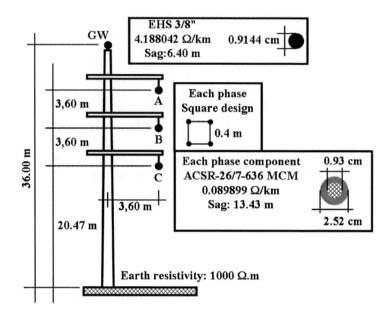

Figure IV.7. Actual asymmetrical vertical three-phase transmission line.

The quasi-mode longitudinal impedance and the quasi-mode shunt admittance matrices have the following structure:

$$Z_{MR} = \begin{bmatrix} Z_{R\alpha} & Z_{\alpha\beta} & Z_{\alpha 0} \\ Z_{\alpha\beta} & Z_{R\beta} & Z_{\beta 0} \\ Z_{\alpha 0} & Z_{\beta 0} & Z_{R0} \end{bmatrix} \quad and \quad Y_{MR} = \begin{bmatrix} Y_{R\alpha} & Y_{\alpha\beta} & Y_{\alpha 0} \\ Y_{\alpha\beta} & Y_{R\beta} & Y_{\beta 0} \\ Y_{\alpha 0} & Y_{\beta 0} & Y_{R0} \end{bmatrix} \quad (90)$$

It is calculated the relative errors of the λ_{IR} and λ_{VR} quasi-modes when compared to the λ eigenvalues using equations (85). These comparisons are shown in Figures IV.8 and IV.9 and the relative errors can be considered negligible. Because of this, the last equations related to the λ_{IR} and λ_{VR} can be rewritten equaling the quasi-modes to the exact eigenvalues and maintaining the coupling among the quasi-modes. This is shown in next equations.

$$\lambda_{IR} = \begin{bmatrix} \lambda_\alpha & \lambda_{IR\alpha\beta} & \lambda_{IR\alpha 0} \\ \lambda_{IR\alpha\beta} & \lambda_\beta & \lambda_{IR\beta 0} \\ \lambda_{IR\alpha 0} & \lambda_{IR\beta 0} & \lambda_0 \end{bmatrix} \quad and \quad \lambda_{VR} = \begin{bmatrix} \lambda_\alpha & \lambda_{VR\alpha\beta} & \lambda_{VR\alpha 0} \\ \lambda_{VR\alpha\beta} & \lambda_\beta & \lambda_{VR\beta 0} \\ \lambda_{VR\alpha 0} & \lambda_{VR\beta 0} & \lambda_0 \end{bmatrix} \quad (91)$$

The rewritten equations are based on the next two figures.

The following two figures are related to the longitudinal impedance and shunt admittance results, respectively

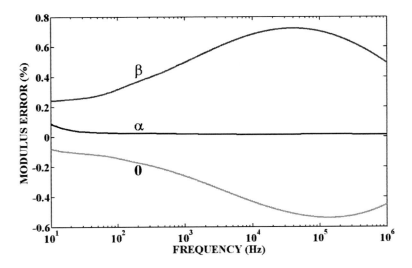

Figure IV.8. Comparisons among λ_{IR} quasi-modes and λ eigenvalues for the actual asymmetrical vertical three-phase transmission line.

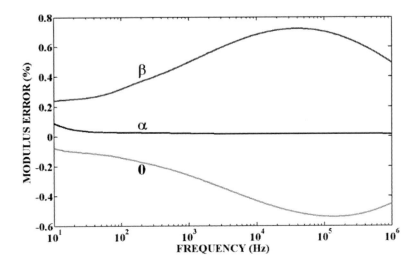

Figure IV.9. Comparisons among λ_{VR} quasi-modes and λ eigenvalues for the actual asymmetrical vertical three-phase transmission line.

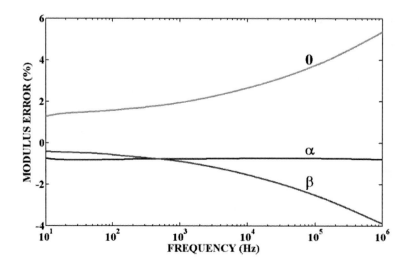

Figure IV.10. Relative errors related to the mode longitudinal impedances for the actual asymmetrical vertical three-phase transmission line.

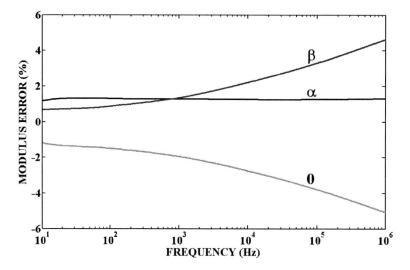

Figure IV.11. Relative errors related to the mode shunt admittances for the actual asymmetrical vertical three-phase transmission line.

The errors of the quasi-mode longitudinal impedances and the quasi-mode shunt admittances reach the highest modulus values in the end of the analyzed frequency range. Depending on the application and the considered frequency, equations (90) can also be rewritten and it is obtained the following:

$$Z_{MR} = \begin{bmatrix} Z_\alpha & Z_{\alpha\beta} & Z_{\alpha 0} \\ Z_{\alpha\beta} & Z_\beta & Z_{\beta 0} \\ Z_{\alpha 0} & Z_{\beta 0} & Z_0 \end{bmatrix} \quad and \quad Y_{MR} = \begin{bmatrix} Y_\alpha & Y_{\alpha\beta} & Y_{\alpha 0} \\ Y_{\alpha\beta} & Y_\beta & Y_{\beta 0} \\ Y_{\alpha 0} & Y_{\beta 0} & Y_0 \end{bmatrix} \qquad (92)$$

It is analyzed the coupling among the quasi-modes. In this case, the next both figures show the $\lambda_{IR\alpha\beta}$ and $\lambda_{VR\alpha\beta}$ in function of frequency values. The $\lambda_{IR\alpha\beta}$ peak value is close to 5 kHz and the $\lambda_{VR\alpha\beta}$ is in the initial of the frequency range. The results related to the T_I eigenvector change are higher than those related to the T_V eigenvector change. So, the $\lambda_{IR\alpha\beta}$ values are in the range from 0.25% to 0.65% while the $\lambda_{VR\alpha\beta}$ values are in the range from 0.04% to 0.18%. This characteristic is true for the other both coupling elements which relative values are shown in Figures IV.14, IV.15, IV.16 and IV.17. In case of Figure IV.14, the peak value is related to frequencies close to 20 Hz and it is higher than that related to the $\alpha 0$ coupling of the symmetrical three-phase transmission line analyzed in the previous item. For Figure IV.15, this peak is related to initial values of the frequency range. The T_I eigenvector change shows higher relative values than the T_V eigenvector change. For the T_I eigenvector change, it is in the range from 0 to 25% while, for T_V eigenvector change, it is in the range from 0 to 2.5%. So, the $\lambda_{IR\alpha 0}$ relative values are about 10 times higher than the $\lambda_{VR\alpha 0}$ ones.

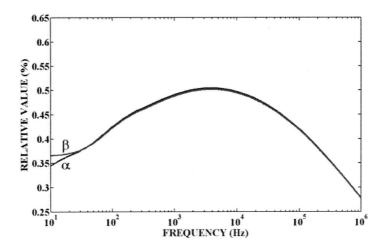

Figure IV.12. The $\lambda_{IR\alpha\beta}$ relative values for the actual asymmetrical vertical three-phase transmission line.

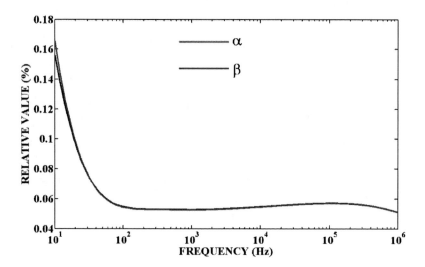

Figure IV.13. The $\lambda_{VR\alpha\beta}$ relative values for the actual asymmetrical vertical three-phase transmission line.

The relative values shown in Figure IV.14 are not negligible values mainly when it is considered the comparisons related to the λ_α eigenvalue. The other non-negligible relative values are obtained for the $\lambda_{IR\beta0}$ element which curves are shown in Figure IV.16.

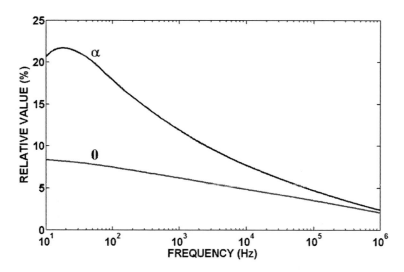

Figure IV.14. The $\lambda_{IR\alpha0}$ relative values for the actual asymmetrical vertical three-phase transmission line.

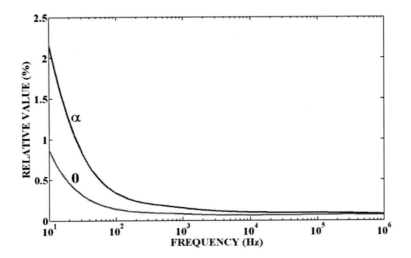

Figure IV.15. The $\lambda_{VR\alpha0}$ relative values for the actual asymmetrical vertical three-phase transmission line.

In this case, the peak value is about 17% and related to frequency values close to 20 Hz. This frequency value is similar to that related to the Figure IV.14.

Besides the peak of 17% for frequency values close to 20 Hz, the range of relative values of the $\lambda_{IR\beta0}$ element is from 4% to 18%. This range is higher than that related to the $\lambda_{VR\beta0}$ element. For the $\lambda_{VR\beta0}$ element, the range of relative values is from 0.5% to 4%. Similar to the other both couplings, the TI eigenvector change presents the highest relative value range in case of Figures IV.16 and IV.17.

The relative values in this item are calculated from the following:

$$\varepsilon_{IRKJ}(\%) = \frac{\lambda_{IRKJ}}{\lambda_K \ or \ \lambda_J} \cdot 100 \quad and \quad \varepsilon_{VKJ}(\%) = \frac{\lambda_{VRKJ}}{\lambda_K \ or \ \lambda_J} \cdot 100$$

(93)

$$K = \alpha, \beta, 0 \quad and \quad J = \alpha, \beta, 0 \quad and \quad K \neq J$$

Analyzing the results shown in the last six figures, the highest relative values are related to the elements that couple the 0 mode to the other both modes. So, considering these highest relative values associated to the $\lambda_{IR\alpha0}$ and $\lambda_{IR\beta0}$ elements, the relative value peaks of each element are associated to the mode different of the 0 mode. It is concluded that the influence of the 0 mode on the other both modes is more significant than the influence of the other both modes on the 0 mode. This is also observed for the typical symmetrical three-phase line investigated in the last item. If this is also observed for another typical asymmetrical three-phase transmission line, it is carried out a correction procedure for Clarke's matrix based on the 0 mode.

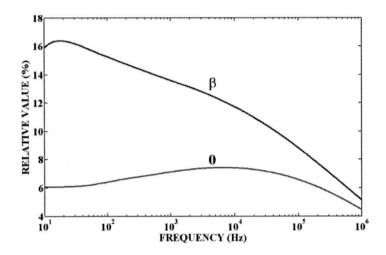

Figure IV.16. The $\lambda_{IR\beta0}$ relative values for the actual asymmetrical vertical three-phase transmission line.

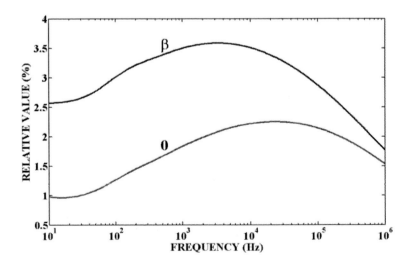

Figure IV.17. The $\lambda_{VR\beta0}$ relative values for the actual asymmetrical vertical three-phase transmission line.

IV.4. UNTRANSPOSED THREE-PHASE LINES WITH PHASE CONDUCTORS DISTRIBUTED IN A TRIANGULAR DESIGN

This line design, called asymmetrical triangular three-phase transmission line, is shown in the next figure. Because there are couplings among all quasi-modes, there are not null values in quasi-mode matrices as well as in longitudinal impedance and shunt admittance matrices in quasi-mode domain.

The relative errors are obtained using equations (85) and the next both figures show the results that depend on the frequency. Because the errors shown in the next both figures are considered negligible, the λ_{IA} and λ_{VA} quasi-mode matrices are described by equations (94) where the quasi-modes are changed into the correspondent λ eigenvalues.

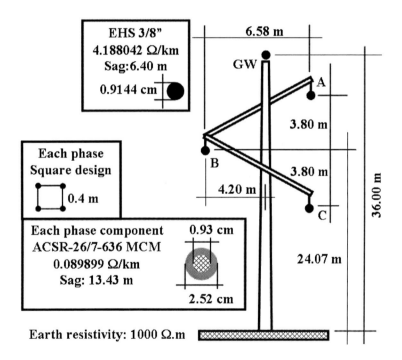

Figure IV.18. Actual asymmetrical triangular three-phase transmission line.

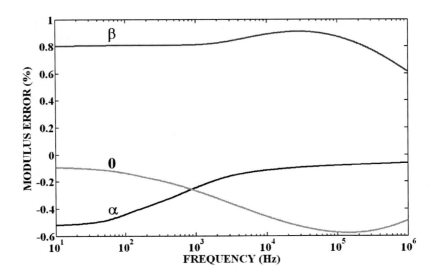

Figure IV.19. Comparisons among $\lambda_{I\Delta}$ quasi-modes and λ eigenvalues for the actual asymmetrical triangular three-phase transmission line.

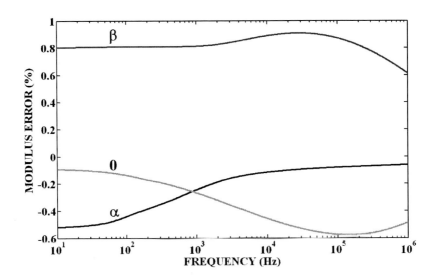

Figure IV.20. Comparisons among $\lambda_{V\Delta}$ quasi-modes and λ eigenvalues for the actual asymmetrical triangular three-phase transmission line.

The $\lambda_{I\Delta}$ and $\lambda_{V\Delta}$ quasi-mode matrices are described by:

$$\lambda_{I\Delta} = \begin{bmatrix} \lambda_\alpha & \lambda_{I\Delta\alpha\beta} & \lambda_{I\Delta\alpha0} \\ \lambda_{I\Delta\alpha\beta} & \lambda_\beta & \lambda_{I\Delta\beta0} \\ \lambda_{I\Delta\alpha0} & \lambda_{I\Delta\beta0} & \lambda_0 \end{bmatrix} \quad and \quad \lambda_{V\Delta} = \begin{bmatrix} \lambda_\alpha & \lambda_{V\Delta\alpha\beta} & \lambda_{V\Delta\alpha0} \\ \lambda_{V\Delta\alpha\beta} & \lambda_\beta & \lambda_{V\Delta\beta0} \\ \lambda_{V\Delta\alpha0} & \lambda_{V\Delta\beta0} & \lambda_0 \end{bmatrix} \quad (94)$$

The impedance and admittance errors are the next both figures.

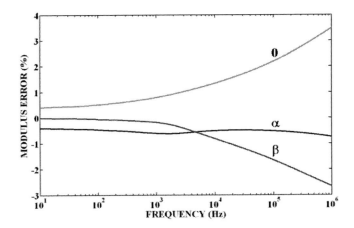

Figure IV.21. Relative errors related to the mode longitudinal impedances for the actual asymmetrical triangular three-phase transmission line.

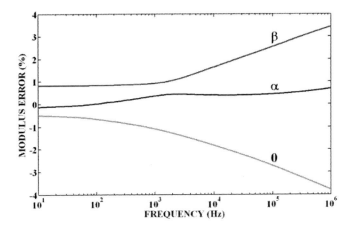

Figure IV.22. Relative errors related to the mode shunt admittances for the actual asymmetrical triangular three-phase transmission line.

The longitudinal impedance and shunt admittance matrices in quasi-mode domain are described by next equations and based on the relative errors shown in the next both figures. The proposed consideration of the quasi-mode values as the mode ones depends on the application and the considered frequency, because the error absolute values in the last both figures are increasing for increasing frequency values.

$$Z_{M\Delta} = \begin{bmatrix} Z_\alpha & Z_{\alpha\beta} & Z_{\alpha 0} \\ Z_{\alpha\beta} & Z_\beta & Z_{\beta 0} \\ Z_{\alpha 0} & Z_{\beta 0} & Z_0 \end{bmatrix} \quad and \quad Y_{M\Delta} = \begin{bmatrix} Y_\alpha & Y_{\alpha\beta} & Y_{\alpha 0} \\ Y_{\alpha\beta} & Y_\beta & Y_{\beta 0} \\ Y_{\alpha 0} & Y_{\beta 0} & Y_0 \end{bmatrix} \tag{95}$$

In case of off-diagonal elements of the $\lambda I\Delta$ and $\lambda V\Delta$ quasi-mode matrices, the relative values are calculated using equations (96) and the results are shown in the next six figures.

$$\varepsilon_{I\Delta KJ}(\%) = \frac{\lambda_{I\Delta KJ}}{\lambda_K \ or \ \lambda_J} \cdot 100 \quad and \quad \varepsilon_{VKJ}(\%) = \frac{\lambda_{V\Delta KJ}}{\lambda_K \ or \ \lambda_J} \cdot 100 \tag{96}$$

$$K = \alpha, \beta, 0 \quad and \quad J = \alpha, \beta, 0 \quad and \quad K \neq J$$

The comparisons of $\lambda_{I\Delta\alpha\beta}$ and $\lambda_{V\Delta\alpha\beta}$ quasi-mode couplings to the correspondent eigenvalues are shown in the next both figures, respectively. The relative values related to the T_I eigenvector change into Clarke's matrix ($\lambda_{I\Delta\alpha\beta}$) are higher than those related to the T_V eigenvector change ($\lambda_{V\Delta\alpha\beta}$). The relative value peak in these both figures is about 0.27% and related to a frequency value about 20 kHz. The relative values shown in these figures can be considered negligible.

Analyzing the $\lambda_{I\Delta\alpha 0}$ and $\lambda_{V\Delta\alpha 0}$ relative values, in Figures IV.25 and IV.26, it is shown results that can also be considered negligible. The relative value peak in these mentioned figures is about 2.4% and it is related a frequency value close to 20 Hz. The quasi-mode coupling relative values based on the YZ product, which is related to the T_I eigenvector, are higher than those based on the ZY matricial product, which is related to the T_V eigenvector. This characteristic is similar to the other analyzed relative values in this item and in the previous both items.

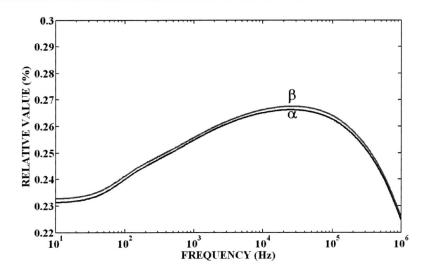

Figure IV.23. The $\lambda_{I\Delta\alpha\beta}$ relative values for the actual asymmetrical triangular three-phase transmission line.

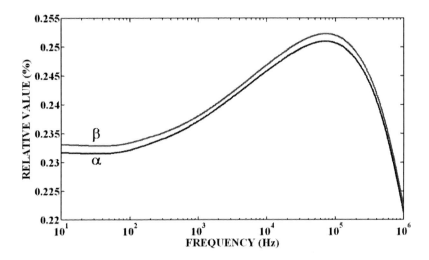

Figure IV.24. The $\lambda_{V\Delta\alpha\beta}$ relative values for the actual asymmetrical triangular three-phase transmission line.

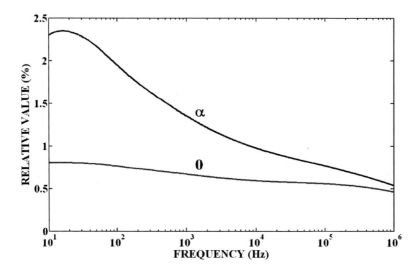

Figure IV.25. The $\lambda_{I\Delta\alpha0}$ relative values for the actual asymmetrical triangular three-phase transmission line.

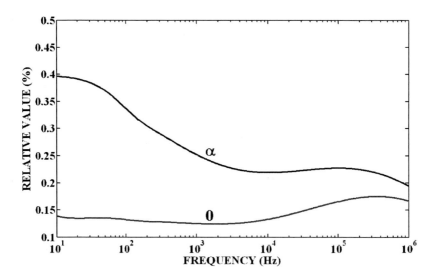

Figure IV.26. The $\lambda_{V\Delta\alpha0}$ relative values for the actual asymmetrical triangular three-phase transmission line.

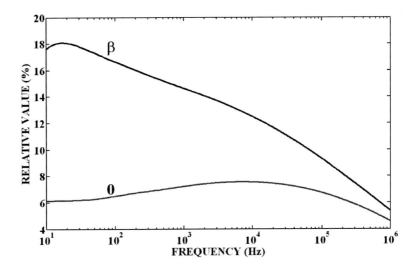

Figure IV.27. The $\lambda_{I\Delta\beta0}$ relative values for the actual asymmetrical triangular three-phase transmission line.

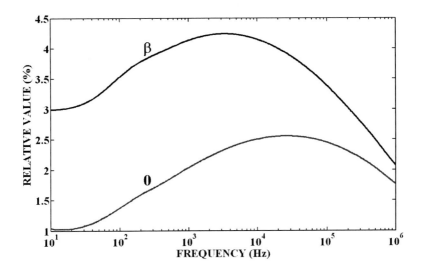

Figure IV.28. The $\lambda_{V\Delta\beta0}$ relative values for the actual asymmetrical triangular three-phase transmission line.

For the asymmetrical triangular three-phase transmission line shown in this chapter, the highest relative values of off-diagonal elements of the quasi-mode matrices are the $\lambda_{I\Delta\beta0}$ ones. These values are shown in Figure IV.27 and the peak value is 18% approximately. This peak is in the initial of the considered frequency range. The Figure IV.28 shows the $\lambda_{V_{\Delta\beta0}}$ relative values, completing the analyses of this item.

The $\lambda_{V_{\Delta\beta0}}$ relative values are lower than the $\lambda_{I_{\Delta\beta0}}$ ones. This is a characteristic property for all relative value couplings among quasi-modes for the analyzed three typical three-phase transmission lines. So, the off-diagonal elements of the quasi-mode matrices obtained from the YZ matricial product have higher relative values than those obtained from ZY matricial product. For untransposed symmetrical and asymmetrical three-phase transmission lines, changing the T_I eigenvector into Clarke's matrix, no negligible couplings among the quasi-modes can be obtained and it can lead to significant errors, if the quasi-mode values are used to calculated other variables in mode domain, neglecting the mentioned couplings. On the other hand, the main advantage associated to the mode domain is the manipulation of diagonal matrices where only the elements of matrix main diagonal are not null. This implies that it is necessary to obtain a better approximation for the T_I eigenvector. An alternative is to apply a correction procedure to Clarke's matrix, improving the approximation for the T_I eigenvector. It is carried out in the next chapter.

Chapter 5

CORRECTION PROCEDURE APPLICATION TO CLARKE'S MATRIX [42]

The correction procedure for Clarke's matrix is analyzed step by step in this chapter, searching for details of each matrix that composes this procedure and considering the typical three-phase transmission lines analyzed previously. Changing the eigenvectors as modal transformation matrices, Clarke's matrix has chosen because this matrix is an eigenvector matrix for transposed symmetrical and asymmetrical three-phase transmission lines. In this chapter, the application of this matrix has been analyzed considering untransposed symmetrical and asymmetrical three-phase transmission lines. In most of these cases, the errors related to the eigenvalues can be considered negligible. It is not true when it is analyzed the elements that are not in main diagonal of the matrix obtained from the Clarke's matrix application. The obtained matrix is correspondent to the eigenvalue matrix and called quasi-mode matrix. Its off-diagonal elements represent couplings among the quasi-modes. So, the off-diagonal quasi-mode element relative values are not negligible when compared to the eigenvalues that correspond to the coupled quasi-modes. Minimizing these relative values, the correction procedure is analyzed in detail.

In mode domain, the frequency influence can be easily introduced on the line parameters. Using phase-mode transformation matrices, all electrical parameters and all line representative matrices are obtained in mode domain [1, 2, 3, 4]. The line representative matrices become diagonal and the frequency influence can independently be introduced for every mode. Applying frequency dependent line parameters also leads to frequency dependent phase-mode transformation matrices. For obtaining voltages and currents in phase domain, it is necessary to use convolution procedures [5, 6, 7, 8, 9, 10 14]. An alternative is to change the exact transformation matrices into single real ones and, then, any values can be determined in phase or mode domain using only a matricial multiplication [3, 11-14].

Modal transformations have been applied to three phase transmission lines using Clarke's matrix based on fact that this matrix is an eigenvector one for transposed symmetrical and asymmetrical three-phase transmission lines. For unstransposed three-phase transmission line cases, when eigenvector matrices are changing into Clarke's matrix, the resulting matrix correspondent to the eigenvalue one is not a diagonal matrix. The result of the Clarke's matrix application are called quasi-mode matrix and the elements of this matrix that are not in the main diagonal represent couplings among the quasi-modes. The off-diagonal quasi-mode relative values are not negligible, if they are compared to the exact modes correspondent to the coupled quasi-modes.

Based on these results, the Clarke's matrix application is analyzed considering symmetrical three-phase lines and a frequency range from 10 Hz to 1 MHz. The quasi-mode errors related to the eigenvalues are studied as well as the off diagonal elements of the quasi-mode matrix. Searching for the off-diagonal element relative value minimization, a perturbation approach corrector matrix is applied to Clarke's matrix. This correction procedure application is analyzed step by step considering the errors related to the eigenvalues and the off-diagonal element relative values.

V.1. THE PERTURBATION APPROACH CORRECTOR MATRIX [5]

The procedure shown in this section is based on a first order perturbation theory approach [3]. This procedure improves the quasi-mode results and obtains a better approximation to the exact values. Initializing, the λ_{IS}, λ_{VS}, λ_{IR}, λ_{VR}, $\lambda_{I\Delta}$, $\lambda_{V\Delta}$ matrices are portioned into two blocks. In this section, these matrices are represented by the λ_P matrix.

$$\lambda_P = \begin{bmatrix} & \lambda_{P22} & \lambda_{P\alpha 0} \\ & & \lambda_{P\beta 0} \\ \lambda_{P0\alpha} & \lambda_{P0\beta} & \lambda_{P0} \end{bmatrix} \qquad (97)$$

The λ_{P22} matrix structure is:

$$\lambda_{P22} = \begin{bmatrix} \lambda_{P\alpha} & \lambda_{P\alpha\beta} \\ \lambda_{P\beta\alpha} & \lambda_{P\beta} \end{bmatrix} \tag{98}$$

Although the line representative matrices are symmetrical, small numeric differences are considered among symmetrical elements of the λ_P matrix. Based on the last two equations, the A_α and A_β elements are determined by:

$$A_\alpha = \frac{\text{trace}(\lambda_{P22}) + \sqrt{\text{trace}^2(\lambda_{P22}) - 4 \cdot \det(\lambda_{P22})}}{2}$$

$$A_\beta = \frac{\text{trace}(\lambda_{P22}) - \sqrt{\text{trace}^2(\lambda_{P22}) - 4 \cdot \det(\lambda_{P22})}}{2} \tag{99}$$

The trace term in the last equation is related to the sum of the diagonal elements of the considered matrix. The det term is the determinant of the considered matrix.

The n_{21} and n_{12} values are determined by:

$$n_{21} = \frac{A_\alpha - \lambda_{P\alpha}}{\lambda_{P\alpha\beta}} \quad and \quad n_{12} = \frac{A_\beta - \lambda_{P\beta}}{\lambda_{P\beta\alpha}} \tag{100}$$

Applying these elements, it is obtained the N_{22} matrix.

$$N_{22} = \begin{bmatrix} 1 & n_{12} \\ n_{21} & 1 \end{bmatrix} \tag{101}$$

Using the N_{22} matrix, it is determined a normalization matrix:

$$N = \begin{bmatrix} N_{22} & 0 \\ 0 & 1 \end{bmatrix} \tag{102}$$

In the next equation, the N matrix is applied to the λ_{VS}, λ_{VR}, $\lambda_{V\Delta}$ matrices matrix for calculating of the A matrix considering the correction applied to the T_V matrix.

$$A = N^{-1} \cdot T_{CL}^T \cdot Z \cdot Y \cdot T_{CL} \cdot N = \begin{bmatrix} A_\alpha & 0 & A_{\alpha 0} \\ 0 & A_\beta & A_{\beta 0} \\ A_{0\alpha} & A_{0\beta} & A_0 \end{bmatrix} \tag{103}$$

Considering λ_{IS}, λ_{IR}, $\lambda_{I\Delta}$ matrices and applying the correction procedure for the T_I matrix correction, the A matrix determination is similar to the T_V one, changing the position of the Z and Y matrices.

$$A = N^{-1} \cdot T_{CL}^T \cdot Y \cdot Z \cdot T_{CL} \cdot N = \begin{bmatrix} A_\alpha & 0 & A_{\alpha 0} \\ 0 & A_\beta & A_{\beta 0} \\ A_{0\alpha} & A_{0\beta} & A_0 \end{bmatrix} \tag{104}$$

The structure of the A matrix is determined from:

$$\lambda = A + \left(\lambda_{CL} \cdot Q - Q \cdot \lambda_{CL} \right) \tag{105}$$

The λ_{CL} is the eigenvalue matrix calculated using Clarke's matrix as an eigenvector matrix and considering the analyzed three-phase transmission line in ideally transposition. This can be obtained with only one coupling value in each Z and Y matrices as well as only one value self value in the mentioned matrices. These unique values can be obtained from the arithmetic media of the correspondent values of the matrices related to the untransposed situation of the considered transmission line.

The last equation leads to:

$$\begin{cases} \lambda_K = A_K, & K = \alpha, \beta, 0 \\ A_{JK} = \left(\lambda_{CLK} - \lambda_{CLJ} \right) \cdot Q_{JK}, & J \neq K \end{cases} \tag{106}$$

The A_α and A_β elements have already in calculated in this section. The last equation is used for calculating the A_0 element, as well as, the Q_{JK} elements. Because the $\lambda_{CL\,\alpha}$ element is equal to the $\lambda_{CL\,\beta}$, this implies that the $A_{\alpha\beta}$, $A_{\beta\alpha}$,

$Q_{\alpha\beta}$ and $Q_{\beta\alpha}$ elements are null. Then, only the Q matrix elements of the third line and the third row are not null. These elements correspond to the 0 mode and are calculated by:

$$Q_{0K} = \frac{A_{0K}}{\lambda_{CLK} - \lambda_{CL0}} \quad and \quad Q_{K0} = \frac{A_{K0}}{\lambda_{CL0} - \lambda_{CLK}}$$

(107)

$$K = \alpha, \beta$$

The Q matrix structure is:

$$Q = \begin{bmatrix} 0 & 0 & Q_{\alpha 0} \\ 0 & 0 & Q_{\beta 0} \\ Q_{0\alpha} & Q_{0\beta} & 0 \end{bmatrix}$$

(108)

The perturbation approach corrector matrix is described by:

$$W = N \cdot (I + Q) \quad and \quad W^{-1} = (I + Q^{-1}) \cdot N^{-1}$$

(109)

The corrected transformation matrices are described by:

$$T_{NV} = W^{-1} \cdot T_{CL}^{T} \quad and \quad T_{NV}^{-1} = T_{CL} \cdot W$$

(110)

V.2. FLOWCHART OF THE CORRECTION PROCEDURE

Analyzing step by step the correction procedure, each matrix that composed the W one is applied and its influence is shown. It is studied the contribution of each W term for minimization of errors and the relative values of the off-diagonal quasi-mode matrix elements. Figure V.1 presents the flowchart for these analyses. Checking the changes into the Clarke's matrix, results carried out for the λ_{NCL} matrix representing an untransposed symmetrical three-phase transmission line [15-17].

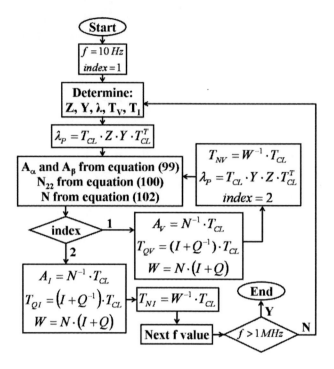

Figure V.1. Correction procedure flowchart.

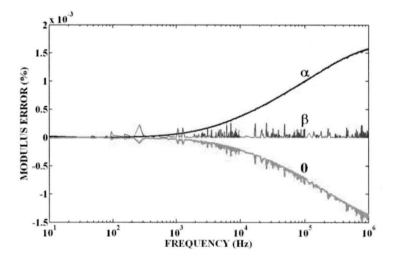

Figure V.2. The quasi-mode relative errors obtained with the Q matrix application for the actual symmetrical three-phase transmission line.

V.3. THE Q MATRIX APPLICATION

Checking the changes into the Clarke's matrix results carried out applying only the Q matrix, it is not used the N and W matrices for corrections based on the T_I and T_V eigenvectors shown in the flowchart of Figure V.1. In this case, only the Q matrix is applied as a correction matrix to Clarke's matrix for checking this matrix influence on the proposed correction. The next both three figures are associated to the Q matrix application to the quasi-modes related to the untransposed symmetrical three-phase line shown in Figure III.4. So, Figure V.2 is associated to the relative errors of the quasi-modes. Figure V.3 shows the relative values of the $\lambda_{IS\alpha0}$ coupling and Figure V.4 is related to the $\lambda_{VS\alpha0}$ relative values. In case of relative errors, the results related to the T_V eigenvector are equal to those related to the T_I eigenvector. Because of this, only one set of curves is shown.

Analyzing only the results related to the symmetrical three-phase transmission line, it is concluded that the Q reduces the quasi-mode relative errors expressively. The decrease is about 150 times when the results obtained for the Q matrix application are compared to those of Figures IV.1 and IV.2. In case of the $\lambda_{IS\alpha0}$ relative values, the decrease of peak value is about 45 times comparing to the peak value of the Figure IV.5. For the $\lambda_{VS\alpha0}$ relative values, it is obtained a 17 time reduction for peak values, approximately, when it is compared the results of Figure V.4 to those related to Clarke's matrix shown in Figure IV.6.

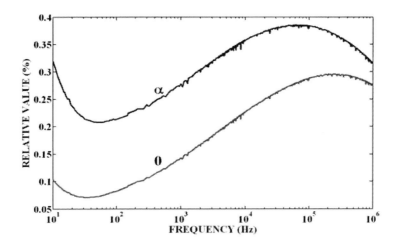

Figure V.3. The $\lambda_{IS\alpha0}$ relative values after the Q matrix application.

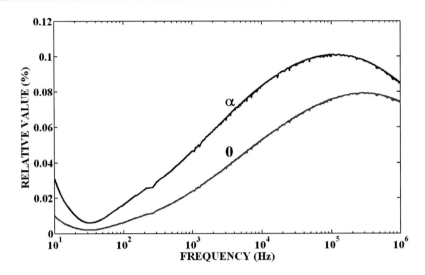

Figure V.4. The $\lambda_{VS\alpha 0}$ relative values after the Q matrix application.

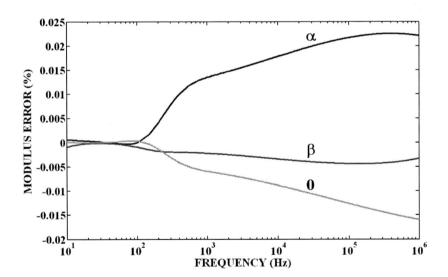

Figure V.5. The quasi-mode relative errors obtained with the Q matrix application for the actual asymmetrical vertical three-phase transmission line.

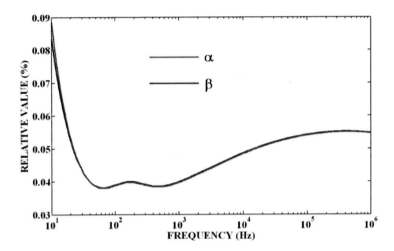

Figure V.6. The $\lambda_{IR\alpha\beta}$ relative values after the Q matrix application.

For asymmetrical vertical three-phase transmission line that is shown in Figure IV.7, after the Q matrix application to correction of Clarke's matrix, the quasi-mode relative error range is decreased about 30 times comparing to Figures IV.8 and IV.9. Besides this, the relative error curve shapes are modified and the β component has values closer to null value than the other components. Previously, this characteristic is associated to the α quasi-mode.

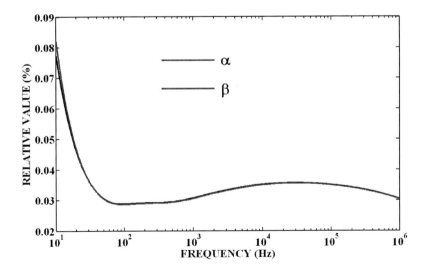

Figure V.7. The $\lambda_{VR\alpha\beta}$ relative values after the Q matrix application.

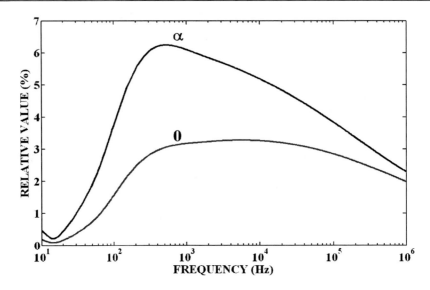

Figure V.8. The $\lambda_{IR\alpha0}$ relative values after the Q matrix application.

Applying the Q matrix and considering the $\lambda_{IR\alpha\beta}$ relative values, it is observed a 5 time decrease in the peak value when it is compared Figures V.6 and IV.12. Besides this, the curve shapes are changed. For the $\lambda_{VR\alpha\beta}$ relative values, the curve shapes are not changed expressively and the reduction of the peak value is about 2 times when it is compared Figure V.7 to Figure IV.13.

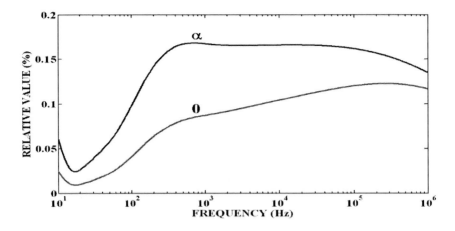

Figure V.9. The $\lambda_{VR\alpha0}$ relative values after the Q matrix application.

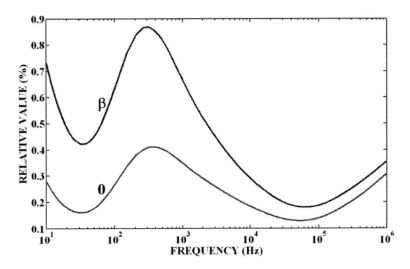

Figure V.10. The $\lambda_{IR\beta0}$ relative values after the Q matrix application.

Analyzing the $\lambda_{IR\alpha0}$ relative values after the Q matrix application in Figure V.8, it is observed that these relative values are decreased about 3.5 times and the curve shapes are modified when compared those curves shown in Figure IV.14. For the $\lambda_{VR\alpha0}$ relative values shown in Figure V.9 when they are compared to those values shown in Figure IV.15, it is noted modifications in the curve shapes and the there is a 10 time reduction in the peak value, approximately.

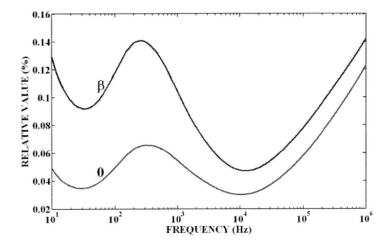

Figure V.11. The $\lambda_{VR\beta0}$ relative values after the Q matrix application.

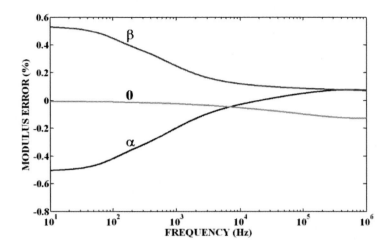

Figure V.12 – The quasi-mode relative errors obtained with the N matrix application for the actual asymmetrical triangular three-phase transmission line.

Considering the $\lambda_{IR\beta 0}$ and $\lambda_{VR\beta 0}$ relative values obtained from the Q matrix application, there are changes in the shapes of the curves comparing Figures V.10 and V.11 to the Figures IV.16 and IV.17, respectively. For the $\lambda_{IR\beta 0}$ relative values, the peak value is reduced from about 17% to 0.9%. It is a 18 time reduction, approximately. Considering the $\lambda_{VR\beta 0}$ relative values, the reduction is about 25 times.

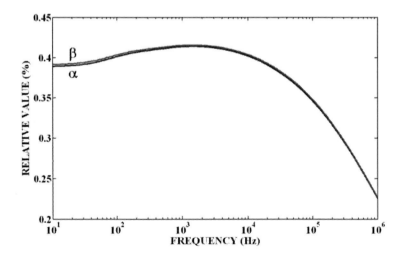

Figure V.13. The $\lambda I \Delta \alpha \beta$ relative values after the Q matrix application.

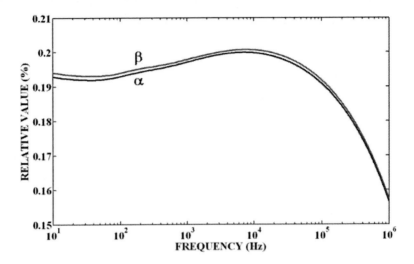

Figure V.14. The $\lambda_{V\Delta\alpha\beta}$ relative values after the Q matrix application.

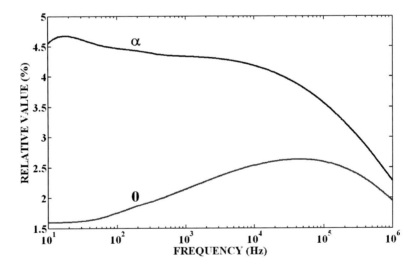

Figure V.15. The $\lambda_{I\Delta\alpha 0}$ relative values after the Q matrix application.

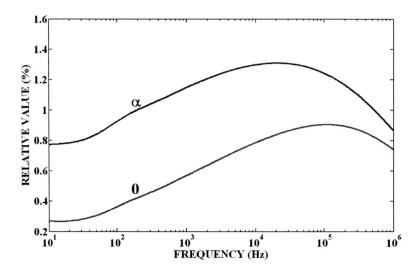

Figure V.16. The $\lambda_{V\Delta\alpha0}$ relative values after the Q matrix application.

For the asymmetrical triangular three-phase transmission line, in Figure V.12, the 0 quasi-mode curve is closer to null value than the same component before the Q matrix application (Figures IV.19 and IV.20). For the other quasi-mode curves, they tend to decrease with the frequency increasing. This characteristic is not observed previously. The modulus error range is not expressively decreased.

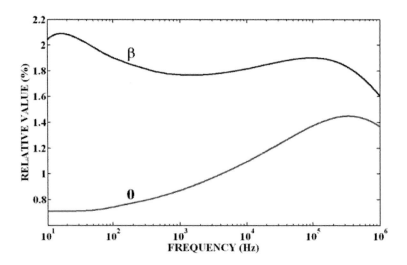

Figure V.17. The $\lambda_{I\Delta\beta0}$ relative values after the Q matrix application.

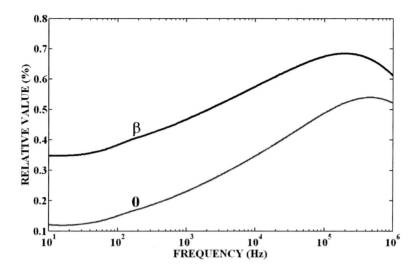

Figure V.18. The $\lambda_{V\Delta\beta0}$ relative values after the Q matrix application.

In Figure V.13, it is shown the $\lambda_{I\Delta\alpha\beta}$ relative values after the Q matrix application. The relative value peak decreases about 1.3 times. Because, in Figure IV.23, these values have already been small, the changes in the values are not significant.

Similar conclusions can be associated to the comparisons between Figure V.14 and Figure IV.24, because the decrease is about 1.2 times for peak value.

For the $\lambda_{I\Delta\alpha0}$ and $\lambda_{V\Delta\alpha0}$ relative values after the Q matrix application, there are changes in the curves shapes, considering the comparisons of Figures V.15 and V.16 to Figures IV.25 and IV.26. While for other couplings related to this analyzed line as well as to the other both lines used in this chapter, for $\lambda_{I\Delta\alpha0}$ and $\lambda_{V\Delta\alpha0}$ relative values, the peak values are increased.

Comparing Figures V.15 and IV.25, it is verified a 1.9 time increasing, approximately, while for $\lambda_{V\Delta\alpha0}$ relative values, this increasing is about 3.2 times comparing Figures V.16 and IV.26.

Considering the $\lambda_{I\Delta\beta0}$ and $\lambda_{I\Delta\beta0}$ relative values after the Q matrix application, there are also changes in the curve shapes and the value peaks are also decreased. For $\lambda_{I\Delta\beta0}$ relative values, it is decreased about from 18% to 2.1%. It is about a 8 time decreasing. In this case, it is compared Figures V.17 and IV.27.

For $\lambda_{V\Delta\beta0}$ values, it is decreased about 6 times, from 4% to 0.7% and it is considering Figures V.18 and IV.28.

The Q matrix application is capable to decrease the all quasi-mode relative modulus errors obtained from the application of Clarke's matrix for all three typical three-phase transmission lines analyzed in this paper. In this case, these lines are considered untransposed. For the off-diagonal element relative values of quasi-mode matrices, the most of these values are reduced with the Q matrix application. Only for the $\lambda_{I\Delta\alpha0}$ and $\lambda_{V\Delta\alpha0}$ values, that are related to the untransposed asymmetrical triangular three-phase transmission line, it is verified that these relative values are increased after the Q matrix application. Based on the shown equations, for completing the correction procedure, it is applied the N matrix. It is analyzed in the next item.

V.4. THE N MATRIX APPLICATION FOR COMPLETING THE CORRECTION PROCEDURE

After the Q matrix use, the correction procedure is completed applying the N matrix. The main function of this matrix is to decrease some coupling values that has had no negligible relative values yet. Considering the symmetrical three-phase transmission line, the relative modulus errors present some differences when they are calculated based on the T_I eigenvectors and the T_V ones. Figure V.19 shows the results associated to the T_I eigenvectors and Figure V.20 shows the results related to the T_V eigenvectors. Comparing the both next figures, there are differences in the curve shapes and in the superior border of the error range. In both figures, the α and β curves present numeric oscillations. For Figure V.19, the decrease of the relative errors is about 35 times when compared to Figure IV.1. Considering Figure V.20, for superior border of vertical axis, the reduction is about 50 times and, for inferior border of this axis, it is about 33 times.

Considering the single coupling for the symmetrical three-phase transmission line and comparing Figures V.21 and IV.5, the $\lambda_{IS\alpha0}$ relative values have a 45 time reduction, approximately. For $\lambda_{VS\alpha0}$ relative values, it is decreased about 17 times when compared Figures V.22 and IV.6. For both figures, V.21 and V.22, the curve shapes are different of those obtained with the application of Clarke's matrix. The frequency values of the peak values are also different, comparing to the results obtained from Clarke's matrix.

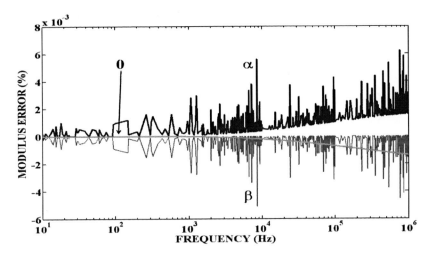

Figure V.19. The λ_{IS} quasi-mode errors after the correction procedure application for the actual symmetrical three-phase transmission line.

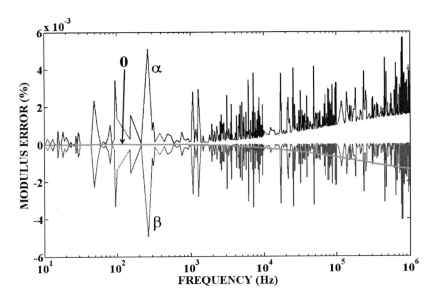

Figure V.20. The λ_{VS} quasi-mode errors after the correction procedure application for the actual symmetrical three-phase transmission line.

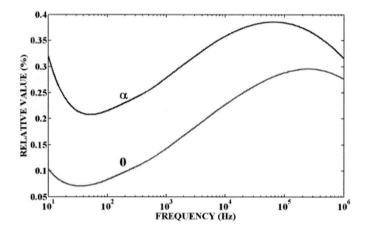

Figure V.21. The $\lambda_{IS\alpha 0}$ relative values after the correction procedure application.

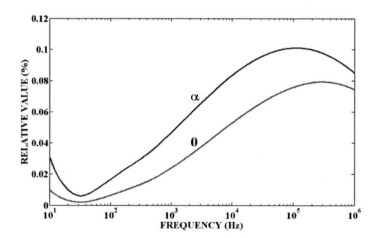

Figure V.22. The $\lambda_{VS\alpha 0}$ relative values after the correction procedure application.

Figure V.23 is related to the asymmetrical vertical three-phase transmission line. Comparing this figure to Figure IV.8, it is observed that there are changes in the curve shapes. The reduction of the relative modulus error range is about 25 times. In this case, there are not differences between the results associated to the T_I eigenvectors and those associated to the T_V eigenvectors.

The results related to the $\alpha\beta$ coupling are shown in Figures V.24 and V.25.

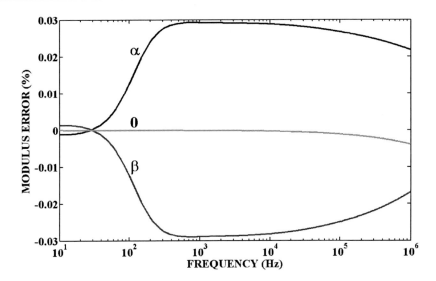

Figure V.23. The relative modulus errors after the correction procedure application for the actual asymmetrical vertical three-phase transmission line.

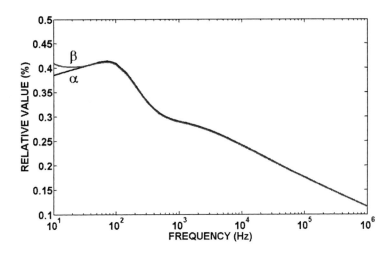

Figure V.24. The $\lambda_{IR\alpha\beta}$ relative values after the correction procedure application.

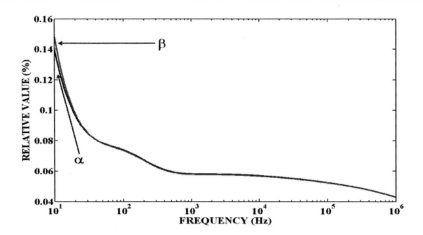

Figure V.25. The $\lambda_{VR\alpha\beta}$ relative values after the correction procedure application.

In Figure V.24, it is observed significant changes only in the curve shapes when compared to Figure IV.12. There is not significant reduction of the relative values. On the other hand, comparing Figures V.25 and IV.13, the changes are not significant for the values and the curve shapes.

The next both figures are related to the $\lambda_{IR\alpha0}$ and $\lambda_{VR\alpha0}$ couplings and they are compared to Figures IV.14 and IV.15, respectively. Only for $\lambda_{IR\alpha0}$ relative values, it is observed significant changes in the curve shapes.

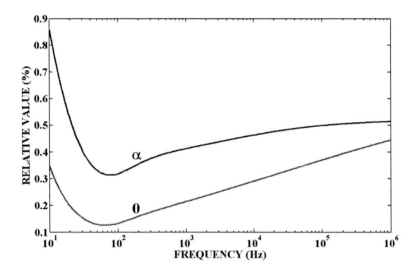

Figure V.26. The $\lambda_{IR\alpha0}$ relative values after the correction procedure application.

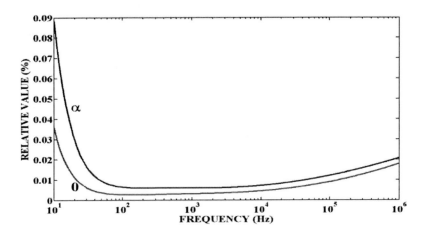

Figure V.27. The $\lambda_{VR\alpha0}$ relative values after the correction procedure application.

Considering the comparisons associated to the $\lambda_{IR\alpha0}$ curves, the reduction is about 25 times, while, for the $\lambda_{VR\alpha0}$ curves, it is about 22 times.

The reduction for the asymmetrical vertical three-phase transmission line, when compared the results obtained from Clarke's matrix and those obtained after the correction procedure application, can be associated to a characteristic value of 25 times. So, comparing Figures V.28 and IV.16, the reduction is about 25 times. In this case, there are also changes in the curve shapes.

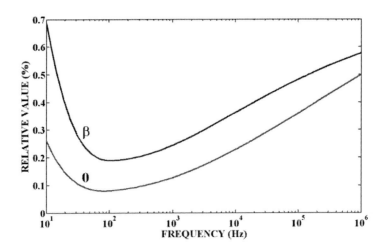

Figure V.28. The $\lambda_{IR\beta0}$ relative values after the correction procedure application.

For the $\lambda_{VR\beta0}$ relative values, comparing Figures V.29 and IV.17, the relative values are decreased about 17 times and there are changes in the curve shapes.

Considering the asymmetrical triangular three-phase transmission line, Figure V.30 shows the relative modulus error curves that is compared to Figure IV.19. With the correction procedure, it is obtained a 2 time reduction for the superior border of the vertical axis, approximately. For the inferior border, the reduction is about 1.25 times.

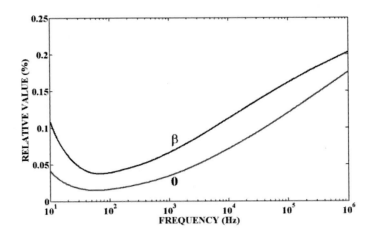

Figure V.29. The $\lambda_{VR\beta0}$ relative values after the correction procedure application.

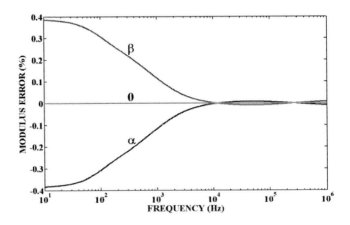

Figure V.30. The relative modulus errors after the correction procedure application for the actual asymmetrical vertical three-phase transmission line.

There are significant changes in curve shapes and in the relative values for high frequencies, when it is compared Figures V.31 and IV.23 as well as Figures V.32 and IV.24. For high frequencies, the reductions of relative values reach about 4.5 times.

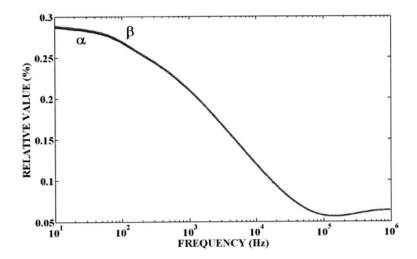

Figure V.31. The $\lambda_{I\Delta\alpha\beta}$ relative values after the correction procedure application.

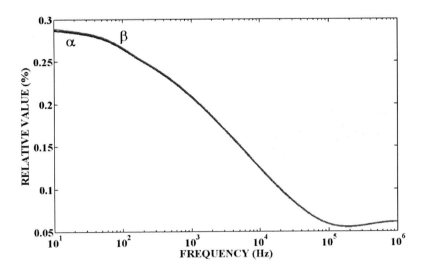

Figure V.32. The $\lambda_{V\Delta\alpha\beta}$ relative values after the correction procedure application.

A 110 time reduction on the peak of the relative values and changes in the curve shapes can be observed comparing Figures V.33 and IV.25. For the $\lambda_{V\Delta\alpha0}$ relative values, it is observed changes in the curve shapes and the reduction reaches about 50 times.

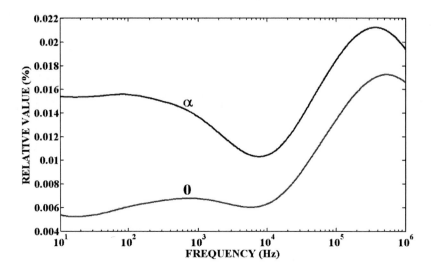

Figure V.33. The $\lambda_{I\Delta\alpha0}$ relative values after the correction procedure application.

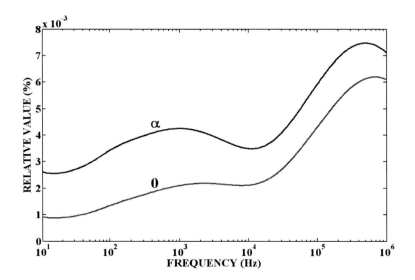

Figure V.34. The $\lambda_{V\Delta\alpha0}$ relative values after the correction procedure application.

For all three typical transmission line samples shown in this chapter, the highest absolute reduction is obtained for the $\lambda_{I\Delta\beta0}$ relative values with the correction procedure application. This reduction is about 140 times. Besides this, there are changes in the curve shapes when Figs V.35 and IV.25 are compared.

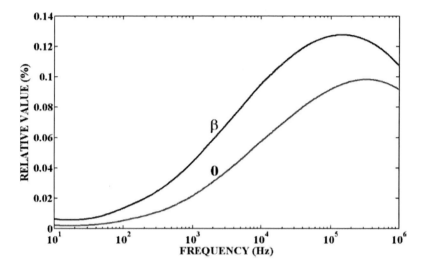

Figure V.35. The $\lambda_{I\Delta\beta0}$ relative values after the correction procedure application.

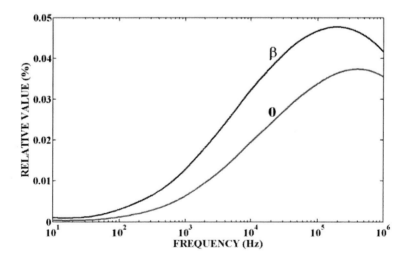

Figure V.36. The $\lambda_{V\Delta\beta0}$ relative values after the correction procedure application.

Finally, considering Figures V.36 and IV.26, it is shown the last comparisons of this item. In this case, the reduction reaches about 90 times and the curve shapes are also modified.

For general analyses, the correction procedure application to Clarke's matrix reduces the relative quasi-mode modulus errors and the off-diagonal relative values of the quasi-mode couplings. With this procedure, it is also obtained balanced relative modulus errors and balanced off-diagonal relative values.

Because the correction procedure is based on Clarke's matrix that is composed by real and constant elements, the both transformations matrices obtained for each analyzed case, one for voltage transformations and the other for the current transformations, are composed by complex elements which imaginary parts are small when compared to their absolute values and real parts. Besides of this, these elements are smoothly influenced by the frequency. These characteristics can be used for creating interesting applications for electromagnetic transient simulations. If the interest is the application of the exact transformation matrices, the obtained matrices from the correction procedure can be used, because the all relative errors and relative values can be considered negligible. Probably, this will be associated to convolution numeric methods. Using the both corrected transformation matrices for a specific frequency, it avoids the convolution numeric methods and frequency scan analyses could be used for validating this application. Simplifying much more, only the real part of the obtained both corrected transformations matrices can be considered. For frequency scan analyses, this alternative has presented negligible differences when compared to the application of the both corrected transformation matrices, considering a symmetrical three-phase transmission line. Based on the obtained results, because the new both transformations matrices can be considered exact transformation matrices, having small imaginary parts and smooth influence of the frequency, it is possible to get some other simplifications from them.

Chapter 6

IMPROVING SPECIFIC RESULTS [39-41]

If the analyzed three-phase transmission line is symmetry or, at least, if it has a phase conductor in the horizontal center of the phase conductors, it is possible to use a modified version of equations (109) and (110), improving some results related to the relative errors and the off-diagonal relative values. In this case, based on the T_V eigenvectors, it is applied the following equations:

$$W_S = N \cdot (I + Q) \cdot N^{-1} \quad and \quad W_S^{-1} = N \cdot (I + Q^{-1}) \cdot N^{-1} \tag{111}$$

Now, the corrected transformation matrices are described by:

$$T_{NVS} = W_S^{-1} \cdot T_{CL}^T \quad and \quad T_{NVS}^{-1} = T_{CL} \cdot W_S \tag{112}$$

Basing on the T_I eigenvectors, it is necessary to change the positions of the Z and Y matrices according to demonstrate in the item V.1 of this chapter. The improved results are the relative modulus errors associated to the symmetrical and asymmetrical three-phase transmission lines. For the asymmetrical vertical three-phase transmission line shown in this chapter, the application of the modified mentioned equations cause more reductions of the $\lambda_{IR\alpha\beta}$ and $\lambda_{VR\alpha\beta}$ relative values without affecting the other ones when compared to the application of the equations (109) and (110).

The next both figures show the improved relative modulus errors related to the symmetrical and asymmetrical vertical three-phase transmission lines, respectively.

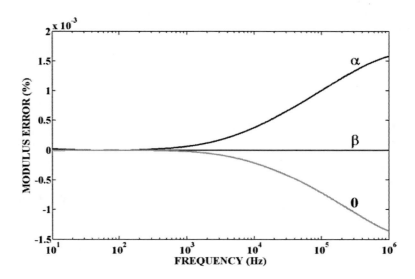

Figure VI.1. The relative modulus errors obtained with the specific correction procedure for the actual symmetrical three-phase transmission line.

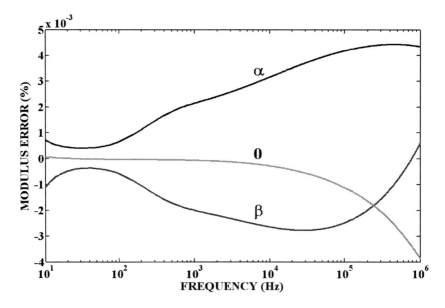

Figure VI.2. The relative modulus errors obtained with the specific correction procedure for the actual asymmetrical vertical three-phase transmission line.

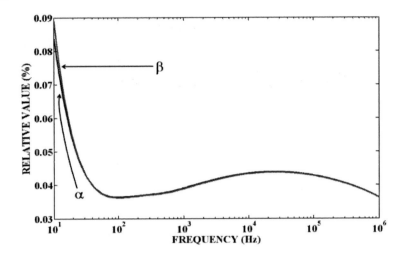

Figure VI.3. The improved $\lambda_{IR\alpha\beta}$ relative values.

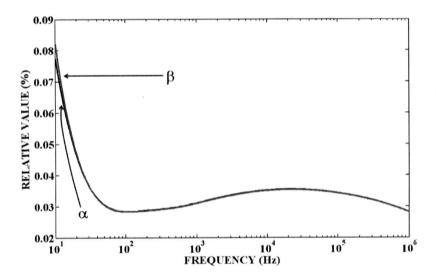

Figure VI.4. The improved $\lambda_{VR\alpha\beta}$ relative values.

Completing the analyses of this item, it is shown in Figure VI.3 and VI.4, the improved $\lambda_{IR\alpha\beta}$ and $\lambda_{VR\alpha\beta}$ relative values, respectively.

Comparing Figure VI.1 to Figures V.19 and V.20, the main changes are that the numeric oscillations are removed. Considering the Figure VI.2, it is obtained a 6 time additional reduction for the peak value when it is compared to Figure V.23. Considering the $\lambda_{IR\alpha\beta}$ and $\lambda_{VR\alpha\beta}$ relative values, the additional reduction reaches about 5 times and 1.75 times for peak values, respectively.

VI.1. OTHER ANALYSES ASSOCIATED TO THE APPLICATIONS SHOWN IN THIS CHAPTER

It has been investigated the inclusion of the technique that reduces the complex elements of the modal transformation matrices in a numeric routine based on state variables. In future, this could be useful for electromagnetic transient simulations [42, 43-45].

It has mentioned that the modal transformation matrices obtained from the proposed correction procedure could be the base for simplified methods of electromagnetic transient analyses and simulations [42]. These simplifications should be checked using frequency scan and error analyses. Time domain analyses and transient simulations could be also applied for investigating the accuracy and the efficiency of the simplified modal transformation matrix applications [39-41].

Modal transformation matrices with constant and real elements have been investigated for application in systems with parallel three-phase circuits [46-47]. The correction procedure should be also adjusted for these systems, extending for different systems the mathematical development applied to three-phase transmission lines in this chapter [42, 44, 46-47].

Chapter 7

CONCLUSIONS

Based on the exact modal transformations, we suggested some applications of this concept for analyses related to transmission lines. We presented an alternative method for calculating the transmission line parameters from impedance values. It can be considered a theoretical method that is the base for the determination of an equivalent conductor from a bundled conductor. This is the second application shown in this book and it used the exact modal transformation matrices for calculating the parameters of the equivalent conductor. For typical symmetrical three-phase transmission line, the exact transformation matrices were separated into two matrices. One of them is a constant real matrix. This technique reduces the complex elements in the obtained transformation matrices, simplifying the numeric routine where these matrices are used.

Considering some approximations for the exact modal transformations, the eigenvector matrices are changed into Clarke's matrix for three typical three-phase transmission line cases. One of these is a symmetrical line and the others are asymmetrical lines. We analyzed the errors of these changes and concluded that the asymmetrical geometrical line characteristics can increase these errors. Because of this, we suggested a correction procedure application to Clarke's matrix. Based on Clarke's matrix, the new transformation matrices have small imaginary parts and their elements are smoothly influenced by the frequency. These characteristics are interesting because from these new transformation matrices, some simplifications can be obtained. For example, only the real part of the new transformation matrices can be used for applications to the EMTP programs.

REFERENCES

[1]. Dommel, HW. *Electromagnetic Transients Program-Rule, Book*, Oregon, 1984.

[2]. Microtran Power System Analysis Corporation, *Transients Analysis Program Reference Manual*, Vancouver, Canada, 1992.

[3]. Morched, A; Gustavsen, B; Tartibi, M. "A Universal Model for Accurate Calculation of Electromagnetic Transients on Overhead Lines and Underground Cables", *IEEE Trans. on Power Delivery*, vol. 14, no. 3, 1032-1038, July 1999.

[4]. Brandão Faria, JA. "Overhead Three-phase Transmission Lines – Non-diagonalizable situations", *IEEE Transactions on Power Delivery*, vol. 3, no. 4, October 1988.

[5]. Brandão Faria, JA; Briceño Mendez, J. "Modal Analysis of Untransposed Bilateral Three-phase Lines - a Perturbation Approach", *IEEE Transactions on Power Delivery*, vol. 12, no. 1, January 1997.

[6]. Brandão Faria, JA. Briceño Mendez, J. "On the Modal Analysis of Asymmetrical Three-phase Transmission Lines using Standard Transformation Matrices", *IEEE Transactions on Power Delivery*, vol. 12, no. 4, October 1997.

[7]. Clarke, E. *Circuit Analysis of AC Power Systems*, vol. I, Wiley, New York, 1950.

[8]. Prado, AJ; Pissolato Filho, J; Kurokawa, S; Bovolato, LF. "Eigenvalue Analyses of Two Parallel Lines using a Single Real Transformation Matrix", *The 2005 IEEE/Power Engineering Society General Meeting*, CD-ROM, 12-16 June 2005, San Francisco, USA.

[9]. Prado, AJ; Pissolato Filho, J; Kurokawa, S; Bovolato, LF. "Non-transposed three-phase line analyses with a single real

transformation matrix", *The 2005 IEEE/Power Engineering Society General Meeting*, CD-ROM, 12-16 June 2005, San Francisco, USA.

[10]. Prado, AJ; Pissolato Filho, J; Kurokawa, S; Bovolato, LF. "Transmission line analyses with a single real transformation matrix - non-symmetrical and non-transposed cases", The 6th Conference on Power Systems Transients (IPST'05), CD-ROM, 19-23 June 2005, Montreal, Canada.

[11]. Wedepohl, LM; Nguyen, HV; Irwin, GD. "Frequency –dependent transformation matrices for untransposed transmission lines using Newton-Raphson method", *IEEE Trans. on Power Systems*, vol. 11, no. 3, 1538-1546, August 1996.

[12]. Nguyen, TT; Chan, HY. "Evaluation of modal transformation matrices for overhead transmission lines and underground cables by optimization method", *IEEE Trans. on Power Delivery*, vol. 17, no. 1, January 2002.

[13]. Nobre, D.M; Boaventura, WC; Neves, WLA. "Phase-Domain Network Equivalents for Electromagnetic Transient Studies", *The 2005 IEEE Power Engineering Society General Meeting*, 12-16 June 2005, CD-ROM, San Francisco, USA.

[14]. Budner, A. "Introduction of Frequency Dependent Transmission Line Parameters into an Electromagnetic Transients Program", *IEEE Trans. on Power Apparatus and Systems*, Vol. PAS-89, 88-97, January 1970.

[15]. Carneiro Jr., S; Martí, JR; Dommel, HW; Barros, HM. "An Efficient Procedure for the Implementation of Corona Models in Electromagnetic Transients Programs", *IEEE Transactions on Power Delivery*, vol. 9, no. 2, April 1994.

[16]. Martins, TFRD; Lima, ACSS. Carneiro Jr., "Effect of Impedance Approximate Formulae on Frequency Dependence Realization", The 2005 *IEEE Power Engineering Society General Meeting*, 12-16 June 2005, CD-ROM, San Francisco, USA.

[17]. Marti, JR. "Accurate modelling of frequency-dependent transmission lines in electromagnetic transients simulations", *IEEE Trans. on PAS*, vol. 101,. 147-155, January 1982.

[18]. Wedepohl, LM. "Application of Matrix Methods to the Solution of Travelling-wave Phenomena in Polyphase Systems", *Proceedings IEE*, vol. 110,. 2200-2212. December, 1963.

[19]. Wedepohl, LM; Wilcox, DJ. "Transient analysis of underground power-transmission system–system model and wave propagation characteristics", *Proceedings of IEE*, vol. 120, no. 2, 253-260, 1973.

[20]. Kurokawa, S; Pissolato, J; Tavares, MC; Portela, CM; Prado, AJ. "A

new procedure to derive transmission-line parameters: applications and restrictions", *IEEE Trans. on Power Delivery*, vol. 21, no. 1, 492-498, January, 2006.

[21]. Hofmann, L. "Series expansions for line series impedances considering different specific resistances, magnetic permeabilities, and dielectric permittivities of conductors, air, and ground", *IEEE Trans. on Power Delivery*, vol. 18, no 2, 564-570, Apr. 2003.

[22]. Portela, C; Tavares, MC. "Modeling, simulation and optimization of transmission lines. Applicability and limitations of some used procedures", *IEEE PES Transmission and Distribution*, 2002, São Paulo, Brazil, 2002.

[23]. Semlyen, A. "Some frequency domain aspects of wave propagation on nonuniform lines", *IEEE Trans. on Power Delivery,* vol. 18, no 1, 315-322, Jan. 2003.

[24]. Akke, M; Biro, T. "Measurements of the frequency-dependent impedance of a thin wire with ground return", *IEEE Trans. on Power Delivery*, (Digital object identifier 101109/TPWRD.2004.834320).

[25]. Koolár, LE; Farzaneh, M. "Vibration of bundled conductors following ice shedding", *IEEE Trans. Power Delivery*, vol. 11, no 2, 2198-2206, April 2008.

[26]. Adams, GE. "An analysis of the radio-interference characteristics of Bundled Conductors", *AIEE Trans. Power Apparatus and Systems*, vol. 75, no.3, pp. 1569-1584, 1957.

[27]. Trinh, NG; Vincent, C. "Bundled-conductors for EHV transmission systems with compressed SF6 insulation", AIEE Trans. Power Apparatus and Systems, vol. 75, no 6, 2198-2206, 1978.

[28]. Dan, VV. "A rational choice of bundle conductors configuration", in *Proc.1998 Int. Symp. on Electrical Insulating Materials Conf.*, Toyohashi, Japan, 349-354.

[29]. Watson, N; Arrilaga, J. *Power Systems Electromagnetic Transients Simulation*, London: Institution of Electrical Engineers, 2003, 140-142.

[30]. Tu, VP; Tlusty, J. "The calculated methods of a frequency-dependent series impedance matrix of overhead transmission lines with a lossy ground for transient analysis problem", in Proc. 2003 Large Engineering Systems Conference on Power Engineering, Montreal, Canada, 159-163.

[31]. Nayak, RN; Sehgal, YK. Sen, S. "EHV transmission line capacity enhancement through increase in surge impedance loading level", in *Proc. 2006 IEEE Power India* Conference, New Delhi.

[32]. Martinez, JA; Gustavsen, B; Durbak, D. "Parameters determination for

modeling system transients - Part I: overhead lines", *IEEE Trans. Power Delivery*, vol. 20, no. 3, 2038-2044, July 2005.

[33]. Mingli, W; Yu, F. "Numerical calculations of internal impedance of solid and tubular cylindrical conductors under large parameters", *IEE Proc. Generation, . Transmission and Distribution*, vol. 151, no 1, 67-72, 2004.

[34]. Tu, VP; Tlusty, J. "The calculated methods of a frequency-dependent series impedance matrix of a overhead transmission line with a lossy ground for transient analysis problem", in *Proc. 2003 Large Engineering Systems Conference on Power Engineering*, Montreal, Canada, 154-158.

[35]. Sinclair, AJ. Ferreira, JA. "Analysis and design of transmission line structures by means of the geometric mean distance", in Proc. 2006 IEEE Africon Conference in Africa, Stellenbosch, South Africa, 1062-1065.

[36]. Kurokawa, S; Costa, ECM; Pissolato, J; Prado, AJ; Bovolato, LF. "An alternative model for bundled conductors considering the distribution of the current among the subconductors", *IEEE Latin America Transactions*, in press.

[37]. Kurokawa, S; Daltin, RS; Prado, AJ; Pissolato, J. "An alternative modal representation of a symmetrical nontransposed three-phase transmission line", *IEEE Trans. on Power Systems*, vol. 22, no. 1, 500-501, February, 2007.

[38]. Fernandes, AB; Neves, LA. "Phase-domain transmission line models considering frequency-dependent transformation matrices", *IEEE Transactions on Power Delivery,* vol. 19, No 2, 708-714, April 2004.

[39]. Prado, AJ; Kurokawa, S; Pissolato, J; Fiho, LF. Bovolato, "Asymmetric transmission line analyses based on a constant transformation matrix", The 2008 *IEEE/PES Transmission and Distribution Conference and Exposition: Latin America*, 13-15 August 2008, CD-ROM, Bogotá, Colombia.

[40]. Prado, AJ; Kurokawa, S; Pissolato Fiho, J; Bovolato, LF. "Single real transformation matrix application for asymmetrical three-phase line transient analyses", *The 2008 IEEE/PES Transmission and Distribution Conference and Exposition: Latin America*, 13-15 August 2008, CD ROM, Bogotá, Colombia.

[41]. Prado, AJ; Pissolato Fiho, J; Kurokawa, S; Bovolato, LF. *"Clarke s matrix correction procedure for non transposed three-phase transmission lines"*. The 2008 IEEE/PES General Meeting, 20-24 July 2008, CD ROM, Pittsburgh, Pennsylvania, USA.

[42]. PRADO, AJ; Kurokawa, S; Pissolato Filho, J; Bovolato, LF. "Voltage and current mode vector analyses of correction procedure application to Clarke's matrix - symmetrical three-phase cases", *Journal of Electromagnetic Analysis and Applications*, vol. 2, no. 1, 12, March, 2010.

[43]. Kurokawa, S; Prado, AJ; Pissolatofilho, J; Bovolato, LF; Daltin, RS. "Alternative proposal for modal representation of a non-transposed three-phase transmission line with a vertical symmetry plane", *IEEE Latin America Transactions*, vol. 7, no. 2, 182-189, June, 2009.

[44]. Kurokawa, S; Yamanaka, FNR; Prado, AJ; Pissolato Filho, J. "Inclusion of the frequency effect in the lumped parameters transmission line model: state space formulation", *Electric Power Systems Research*, vol. 79, no. 7, 1155-1163, July, 2009.

[45]. Kurokawa, S; Daltin, RS; Prado, AJ; Pissolato Filho, J. "An alternative modal representation of a symmetrical non-transposed three-phase transmission line", *IEEE Transactions on Power Systems*, vol. 22, no. 1, 500-501, February, 2007.

[46]. Campos, JCC; Pissolato Filho, J; Prado, AJ; Kurokawa, S. "Single Real Transformation Matrices Applied to Double Three-phase Transmission Lines", *Electric Power Systems Research*, vol. 78, no. 10. 1719-1725, October, 2008.

[47]. Prado, AJ; Pissolato Filho, J; Kurokawa, S; Bovolato, LF. "Modal transformation analyses for double three-phase transmission lines", *IEEE Transactions on Power Delivery*, vol. 22, no. 3, 1926-1936, July, 2007.

INDEX